Cultural Collisions

Cultural Collisions
Postmodern Technoscience

Raphael Sassower

ROUTLEDGE
New York and London

Published in 1995 by
Routledge
29 West 35th Street
New York, NY 10001

Published in Great Britain by
Routledge
11 New Fetter Lane
London EC4P 4EE

Printed in the United States of America on acid-free paper.

Library of Congress Cataloging-in-Publication Data

Sassower, Raphael.
Cultural collisions : postmodern technoscience / Raphael Sassower.
p. cm.
Includes bibliographical references and index.
ISBN 0-415-91109-5.—ISBN 0-415-91110-9 (pbk.)
1. Technology—Social aspects. I. Title.
T14.5.S27 1995
303.48'3—dc20 95-24651
CIP

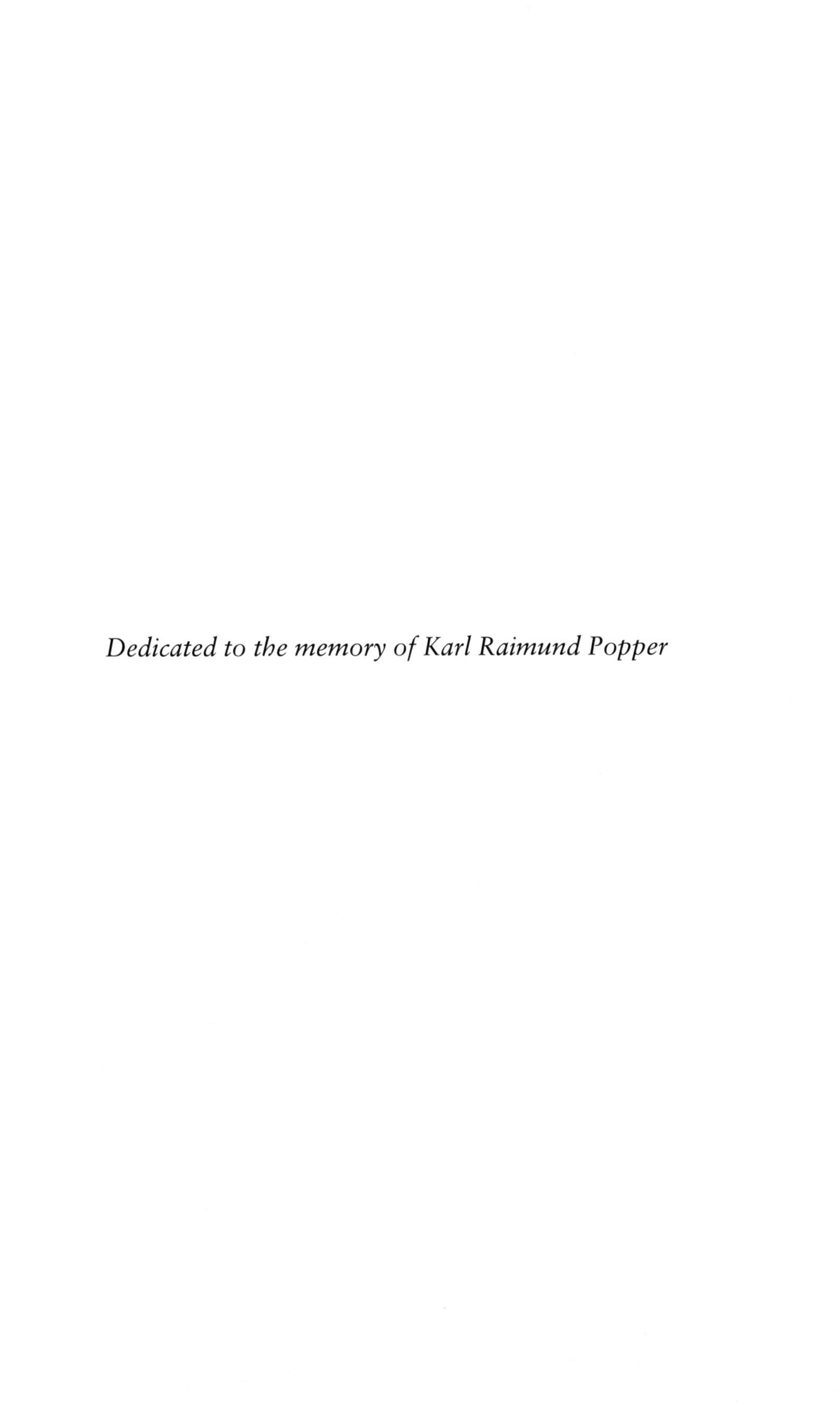

Dedicated to the memory of Karl Raimund Popper

Table of Contents

Preface

My concern with epistemological issues is a political concern that has a psychological component, whether that component is understood socially or personally. The political concern is not that of consensus-building, but of survival, sometimes around a negotiation table, sometimes in a battlefield. In *Knowledge Without Expertise* (1993a) I concluded with poker playing and bluffing as ways of empowering disenfranchised political groups and of bringing them to a round table; I wish to begin here where I left there. The promise of bluffing as a means to overcome entrenched power relations may lead to unexpected complications whereby there is no change in the distribution of power. In its stead, I have come to realize that perhaps one needs to think more carefully about the necessity, not merely the possibility, of pursuing a conversation among groups whose official difference puts them at odds with each other. While one may enter a poker game in any town, one cannot always leave the poker table at will.

If epistemological concerns are indeed political, and the politics is that of conversation, however benign conversations may be deemed at first sight or however crucial they may turn out to be, then my concern here is with the conditions that make it necessary, not possible, to have a conversation. My concern, though, deviates from Kant's Enlightenment Project and its Habermasian afterthought, because I relinquish the dream of consensus-building. Moreover, the conditions for the very possibility of dialogue can be construed as excessive and thereby oppressive: for example, why must we all submit to rationality and its peculiar rules of debate? From the perspective of the necessity of the dialogue, the question is: what is at stake if we refuse to speak with one another, if we leave the poker table? And here political conflicts of this century, some of which are still ongoing, teach us at least one important lesson: human lives are at stake.

People die when their leaders and representatives refuse to talk to the enemy. Subversive activities of underground movements, such as bombing buses and train stations, have grave consequences whose justification must be proven beyond the shadow of the doubt, similar to strict legal tests for finding someone guilty or innocent in cases that warrant capital punishment. Before human life can be offered for sacrifice in the name of justice and revolution or in the name of fairness and dignity, honor, and freedom, it must be absolutely clear that no alternative mode of action, such as negotiation, could yield any reasonable results.

Once it is believed (it is belief and perception of which I speak by necessity, perhaps because of the ties I see in the relation between epistemology and psychology) that some form of communication is needed to prevent bloodshed, then the following question may arise: what rules can be agreed upon? Here I prefer to offer the most simple threshold of intellectual exchange and not an ideal speech-communication. For example, disputants can agree on the common-sense principles of identity, non-contradiction, and the excluded middle, the three basic logical principles used for generations (since Aristotelian syllogisms) to prove the existence of God as well as to undermine divine authority. I hasten to add, though, that there may be reasons to jettison these principles—for example, the principle of non-contradiction—when it becomes useful to present multiple interpretations, some of which contradict each other. Perhaps it would also be advantageous to have the disputants, as in the case of the Israelis and the Palestinians, speak a language foreign to them both, English, rather than either side having the advantage of its native language, Hebrew or Arabic. Moreover, the use of another language opens up possible interpretations that may turn out to be useful for the process of negotiation, so that no statement is final in the sense of being absolutely transparent.

Of course, there are situations when the translation from one language to another does not produce a multiplicity of credible interpretations, but an immediate need to choose an appropriate one. So the multiplicity that allows ambiguity and choice, and from which a negotiation is launched, may at times turn out to be a liability. But even when a single interpretation is chosen, it can be challenged and exchanged for another, for the disclosure of its implication may require a reconsideration. Purposeful evasiveness in this context would ensure an ongoing dialogue in which clarity is achieved over time and is not the precondition for the process of negotiation to be launched.

All of this is a response to the charge that any and all communication and dialogue, whether that of poker players or combatants, closes the discussion or avoids the cultural collisions implied therein by bringing people who disagree with each other together for the sole purpose of reaching an agreement. This theme runs throughout this book in a dialectical tension: wishing to find a common ground where it can be found, while acknowledging the undesirability of such a goal. As far as I am concerned, no agreement may be reached and people will still die in the streets—by military forces or by hunger. At times this horrible fate will be averted in the name of temporary agreement, but one whose temporality is fragile because it is not guaranteed by a permanent consensus. For consensus is neither the goal nor the ideal some portray it to be; for me it is a feared abyss one should avoid, because consensus bespeaks harmony and homogeneity, conditions inconceivable in the postmodern age of difference and fragmentation. Besides, in the name of consensus too much may be given up by participants, so that eventual resentment and frustration will undermine it. My preference is for a dialogue, a conversation—in the form of cultural collisions—in the present context, one between postmodernism and technoscience, however broadly or narrowly these categories are defined.

What, then, is *my* role in the context of conversations as modes of cultural collisions? I am a foreigner and a translator, delivering the messages of others in my own peculiar way so that "others" may hear news not delivered through their regular channels of communication. This requires me to be a recipient of messages as well, one who decodes and interprets, who oversimplifies in the name of comprehension, and who delivers broad, general messages and not the details of discourses, without thereby setting up hierarchies of discourses or meanings. As a foreigner I am not an ambassador, for there is no "official" knowledge to impart, only partial knowledge, the kind one gathers at a meeting, the kind that inspires and excites one to wake up in the middle of the night with a thought, image, or idea.

As a translator I must be kind and generous, even soft-spoken when I am outraged, for as a foreigner I am always a guest, someone whose very presence is conditioned by the hospitality and generosity of others. I walk into a European coffee house looking for an empty table but only find empty seats. I sit down at a table whose other seats are occupied. People stare, say something, and continue their conversation. In America I have to wait for an empty table, for a chair is not enough; if I cannot have the entire table, then I cannot have a chair. Is it a question of privacy that shifts the focus of attention to the entire table, while in Europe the table serves more as a public site and only the chairs are associated with one's privacy? Or is it a question of having the entire stage for oneself in order to perform at all?

When I sit down at a European coffee house, there are other people seated around the table, having their own conversation. Should I eavesdrop? Should I engage someone? Should I insist that we all participate in the same conversation? Should we all speak English or German or French or Spanish or Italian? Must we all drink coffee, or can we order beer or wine, breakfast or lunch or dinner? As I approach an outdoor cafe, these questions whirl in my mind, all of which contain an implied invitation that seems to have no constraints: multiple options present themselves.

Perhaps these questions, now contextualized within the bourgeois comforts of coffee houses rather than Western-style poker tables or the battlefields of the Middle East, explain my concern with translation among different discourses: are all discourses equally privileged? Do they all deserve the same attention? Must they be translatable in principle or in practice? Will their difference be compromised by translation? The concern is postmodern in nature, because the open-endedness of post-modernity embodies the difficulty of not having clear answers or having too many answers from which to choose.

Modernity fought against despotic injustice and religious dogma, and turned out to be at times unjust and dogmatic. Postmodernity, in order to legitimate its militancy and its longevity, has its villains too. Perhaps it fights against the injustice and terror of rationality and the dogmas of science, while being at times both rational and dogmatic. To speak, therefore, about postmodern technoscience, as I try in this book, is to speak dialectically. This speech may reflect some forms of cultural collision that, in order to diffuse their pain and ugliness, require conversations among a variety of participants who are committed to different discourses and language games.

Whether the perspective from which this book is read is that of combatants in war or that of academics, as a translator I wish to request one thing: let me hear the participants, for otherwise my translation would be impossible. In other words, if, in the name of difference, wholesale slaughter is sanctioned, I will hear no one. For example, before prejudging the horrible consequences of laser technology, especially those commonly associated with missile technology, there may be cause to pause and imagine the usefulness of laser technology for medical surgery. This move, I am aware, may be deemed by some as an excuse to legitimate any and all technoscientific research and development. According to this mode of justification, eventually there is a so-called peace dividend of turning spears into ploughshares. For me, though, this move is a way to ensure that our imagination keeps up with technoscientific developments, as I argue at the conclusion of the book. Perhaps a postmodern orientation may be of help to accomplish this task, and perhaps the accomplishment of this task will bring us closer to dealing with cultural collisions in ways that will not set up the conditions that silence all too many groups and usher in another world war.

I refrain from defining here either what postmodern technoscience is or does, because the burden of the entire book is to make some sense of these terms, bring them into collision, translate them from and into different discourses, and explain why it is useful to consider cultural collisions and their consequences. So, instead of writing a brief and inappropriate definition here of either postmodernism or technoscience, I have written an entire book. Because this book has some underlying themes that reappear in every chapter and has specific topics of discussion in each chapter, the order in which it is read depends entirely on the readers and the burning questions with which they approach it. Rather than summarize each chapter in turn, I will try to explain the rationale for my structure. Despite popular belief to the contrary, I do take full responsibility for my authorship and my agency, writing as I do in the postmodern era of technoscience. This is how I perceive in most general terms the structure of the book and its seven chapters.

In Chapter 1, I begin with the story of the Superconducting Supercollider, juxtapose it to some scientific and political concerns of the Weimar Republic, and maintain that public projects, even those deemed technoscientific in nature, cannot live by their scientific credibility alone; they need economic and political support for their very survival. I continue to illustrate what the cultural context of postmodern technoscience may look like in Chapter 2. I survey some common concerns relating to postmodernism and the related concerns with philosophy of science by the end of the twentieth century. In Chapter 3, I present the need for translation as a form of dealing with the inevitability of cultural collisions, understood in the present context in terms of discourses and disciplines. I continue, in Chapters 4 and 5, to give particular examples of discursive translations, between some Popperians and some postmodernists, and between some Popperians and some feminists. In both cases, though, it is assumed that we are dealing with ideal types and not with a generalization from which particular conclusions can be drawn concerning all participants. I also illustrate to what extent there can be a useful exchange of ideas and insights between different discourses, without thereby diminishing either the difference of these discourses or their divergent political commitments.

Chapter 6 is concerned with the inevitability of reality as a shared concern of even the staunchest of critics of science, for there is some agreement that some reference must be made to the reality both natural and social that surrounds us all. I continue to examine the material conditions (of political economy) under which the technoscientific projects in the postmodern age may be more effectively undertaken. I conclude the book in Chapter 7 by discussing why art and imagination are significant pedagogically and bring about a more responsible way of engaging the kind of cultural collisions we are bound to face in the next century.

I call for intellectual responsibility and for a more conscious effort on the part of academics to ensure our survival and the survival of ecological systems within which we have found refuge.

* * *

It is a pleasure to acknowledge individuals whose critical reading helped me as I proceeded writing this book. Among them are Thornton

Tibbals, Charla Phyllis Ogaz, Sam Gill (Chapter 6), Joseph Agassi (Chapter 4), Cheryl Cole (Chapter 5), Phyllis Rooney (Chapter 5), David Whitcombe (Chapter 1), Jay Buick, and Susan Britt. I wish to thank Stanley Aronowitz for alerting me to the work of Paul Forman (for Chapter 1), and Paul van Dijk for bringing to my attention Gunther Anders' work (for Chapter 7). Special gratitude goes to Julia Hoerner, whose artwork "Who Made Thee? (Object Mode)" of 1988 adorns the front cover, and whose conversations with me informed my conception of technoscientific art (Chapter 7).

I must also acknowledge the following journals for their permission to reprint my work. "Postmodern Philosophy of Science: A Critical Engagement," previously published in the *Journal of the Social Sciences* (Canada, 1993), forms part of Chapter 4. "The Politics of Situating Knowledge: An Exercise in Social Epistemology," previously published in *Argumentation* (The Netherlands, 1994), forms part of Chapter 5. "Verklighetens ofrankomlighet" ["The Inevitability of Reality"], previously published in *Vest* (Sweden, 1994), forms part of Chapter 6. I am grateful that intellectual property rights permit us to reclaim our work for ourselves.

Chapter 1

The Superconducting Supercollider

Introduction

As subsequent chapters will illustrate, there is a certain inevitability to the ontology about which even postmodern technoscientists have to speak, as well as to the material conditions under which its performances are observed. Whether we continue the Marxist tradition and speak of political economy in scientific terms, or adopt a postmodern orientation that discards the necessity of using a scientific vocabulary to legitimate a discourse (and a practice), we may wish to examine more carefully what some of my more general comments may mean in the political arena of late twentieth-century America. It is in this light, then, that I recommend examining the fate of the Superconducting Supercollider, a project whose genealogy and eventual demise may inform us about the ideological backdrop against which policy decisions are made. Moreover, I believe that this project sheds some new light on the debate concerning Big Science whose own genealogy is intimately connected in America to the Manhattan Project, the development of the atomic bomb in Los Alamos, New Mexico. Finally, this chapter also intimates what will appear at the closing of this book (Chapter 7) where I try to argue that the artistic nature of postmodern technoscience puts a different set of burdens and responsibilities on intellectuals in general and academics in particular.

Let me begin, as so many nowadays are fond of beginning their discursive explorations, not with Marx (whose texts are less frequently quoted than they were two decades ago) but with Nietzsche (who is presumed to be responsible for everything worthwhile in the intellectual world):

> Do you really believe that the sciences would ever have originated and grown if the way had not been prepared by magicians, alchemists, astrologers, and witches whose promises and pretensions first had to create a thirst, a hunger, a taste for *hidden* and *forbidden* powers? Indeed,

> infinitely more had to be *promised* than could ever be fulfilled in order that anything at all might be fulfilled in the realm of knowledge. (Nietzsche 1974, 240)

I partially examine (in Chapter 3) Nietzsche's claim concerning the need for magicians and alchemists, astrologers and witches, not only in the sense understood in the history of science (e.g., by Newton) or in the sense of oracles and prophets (Sassower 1993a, Ch. 2), but also in the Romantic sense of the past century. His contention applies equally well to the contemporary culture of technological feats. My sense is that postmodern technoscience, however defined in the following chapters, is a way of addressing Nietzsche's contention about the discrepancy between science's promise of discovering hidden treasures and the fulfillment of this promise. This leads me directly to public perceptions, political policy-making processes, and the eventual confrontation between the discourses and communities of science and politics (Sassower 1993a, Ch. 5).

Postmodern technoscience is neither an assemblage of the various critiques of science and technology that have emerged in the late twentieth century, nor an attempt to provide a chronological break between modernism and a successor age. Instead, it is an attempt to introduce and infuse the study and practice of technoscience with a sense of the attitudes and orientations commonly perceived to be underlying the works of postmodernists (however broadly defined), and perhaps, following Bruno Latour, that of premoderns (more on this in Chapters 2–4). Among these attitudes it may be important to inject an inspired use of the imagination in addition to more common ones, such as the multiplicity of interpretations, the deprivilegizing of all discourses, and the contextualization of judgments to particular situations without an appeal to universal and permanent foundations.

Once understood culturally and contextualized into specific situations and conditions, the activities that come under the label of technoscience can be appreciated in their postmodern garb, one that refers to them as cultural artifacts with no more nor less validity and credibility than any other forms of expression and modes of behavior. As such, then, postmodern technoscientific activities are open to public scrutiny and debate in ways that may have been inconceivable only a few decades ago.

For example, the *New York Times* reported in July of 1992 that Congress might not approve in the 1993 fiscal year $8.25 billion for the Superconducting Supercollider in Texas (Browne 1992). How should this piece of technoscientific information be approached? Is it necessary to contextualize this piece of economic information in the terms suggested by physicists who are supporters of the project, some of whom promise great discoveries, or in the terms of other physicists who claim that such an expenditure is unwarranted? Should we compare this figure to some other budgetary figures for the same fiscal year? What warrant should we afford to the rhetorical expressions sounded in some public quarters? I shall return to this case in the next section.

Here are some figures worth noting: The National Endowment for the Arts was supposed to have a budget of $176 million for 1993; the National Science Foundation $2.3 billion; Occupational Safety and Health Review Commission $7.2 million; Office of Government Ethics $8.6 million; and the entire Department of Education $6.8 billion. I have selected these numbers for comparison because they are closely related to public perception and interest as they were directed toward this technoscientific research. They do not compare the Supercollider project with an Air Force bomber or other Department of Defense expenditures, even though they are always related, at least since World War II, to scientific research and development. Should a select group of physicists, one of eighteen identifiable sub-fields in physics, receive support that is almost fourfold the support of the entire scientific community?

Back to Nietzsche. According to Nietzsche, there have been three reasons—for him they constitute three sets of errors—for the promotion of science: first, "it was by means of science that one hoped to understand God's goodness and wisdom best"; second, "one believed in the absolute utility of knowledge"; and three, "one thought that in science one possessed and loved something unselfish, harmless, self-sufficient, and truly innocent" (Nietzsche 1974, 105–106). Science displaces the social and ideological role of religion at least as a practice that craves divine revelation, assumes it to be beneficial for human beings here and in the afterlife, and claims to bring together individuals into a community of inquirers.

Without mentioning Nietzsche, Robert Merton in fact concurs with these reasons for the appeal of science to society, showing how science in seventeenth-century England was supported by the Puritans, how there was a great deal of engineering interest in scientific research, and how the notion of human curiosity and the search for Truth have driven western civilization to new horizons of national support for science and technology (Merton 1968, IV). A similar sentiment is echoed by the "gentlemen of science" in nineteenth-century Britain, one that gave rise to the British Association for the Advancement of Science and to the establishment of an intimate link between political institutions and the scientific community (Morrell & Thackray 1981).

Has the atomic bomb changed our optimistic views of technoscience? Has the destruction of the environment brought us closer to the pessimism, sometimes confused with nihilism, so often associated with postmodernism? That is, a pessimism that is Romantic in nature (that of Rousseau or of Nietzsche) and is nothing more nor less than the flip side of the Enlightenment's optimism? To some extent, and in the spirit of postmodernism, it may be useful to supersede the binary opposition set up between optimism and pessimism and between the Enlightenment and Romanticism respectively. Perhaps what remains at stake is the rhetoric of promise and glory, of future adventures in the name of science, that overlooks the technological apparatus with which this promise must be fulfilled. This is not to say, as some would say about the development of nuclear energy, that the very ideas of science are bound to inflict the horrors of their technological application. On the contrary, there is no science without technology, so that the very conceptions of scientific theories are intimately connected to technological conditions (Ormiston & Sassower 1989, Ch. 1). Post World War II, there is a postmodern technoscientific world that is a "two-faced Janus," as Latour calls it (Latour 1987), one whose optimism is laced with pessimism so that neither disposition ever overcomes the other.

Even traditional philosophers of science, such as Stephen Toulmin, have become aware of the impracticability of holding onto some particular pre-World War II binary oppositions. As I will discuss in the next chapter, though trying to denote "post-modern" science in chronological terms, Toulmin is sensitive to the implications of transforming the human relation with the world from that of a "spectator" to that of a "participant" (Toulmin 1981). The participation that Toulmin recognizes is analogous to that of Latour (1986), for both appreciate the concerns of the "fathers" of quantum mechanics such that laboratory experiments and measurements interrupt the activities of natural phenomena so that instead of "discovering" something out there, scientists in fact "create" or "establish" something right here.

The Superconducting Supercollider

The first report in the *New York Times*, a bellwether of American perceptions of political trends and investigative reporting pitched to the middle and upper middle class, appeared in September 1983, under the provocative title: "Physicists Compete for the Biggest Project of All." The Nietzschean promise appears in the opening sentence: "Put simply, the project would be the biggest endeavor in the history of pure science, a colossus that would rival the building of the pyramids and the construction of the Panama Canal." It was reported that the projection is that a "circular tunnel will stretch anywhere from 60 to 120 miles," and that "its total cost might run from $2 billion to $4 billion." (Broad 1983)

What President would not wish to become a Pharaoh, especially for the low cost estimated at that point?

At the end of January 1987, the *New York Times* reported that the Reagan Administration announced its intentions to ask Congress for about $6 billion to build a giant "atom smasher," a much higher price-tag than originally estimated in 1983. The front-page article continued in the following manner:

> The device, a Superconducting Supercollider in a 52-mile oval tunnel, would dwarf existing machines used to probe the secrets of matter and energy. [in 1983 it was 60 to 120 miles.]
>
> The project to build the world's largest research machine, in which subatomic particles moving at high speed would collide and burst, is as scientifically significant as America's 1969 manned landing on the moon, Secretary of Energy John S. Harrington said. (Franklin 1987)

The rhetoric has not changed since 1983: largest machine in the world (biggest in 1983), dwarf other research devices, comparison with the manned spaceship in 1969 (pyramids and the Panama Canal in 1983), and the probing of the "secrets of matter and energy." What else could one do for the sake of good old-fashioned patriotism? The Secretary of the Department of Energy became a spokesperson for the $6 billion in funding, justified in part by the need to be on the cutting edge of science and, as we read a bit later, by the creation of 4,500 jobs and a staff of at least 2,500 scientists. So it made scientific and economic sense to embark on this project, at least for those Keynesian New-Dealers who knew that public projects stimulate the economy by providing jobs that lead to increased consumption which leads to increased production.

Yet even the rhetoric of a cabinet member did not silence all criticism. Some critics acknowledged that the project would necessarily divert funds from "less glamorous, but equally important," areas of research, while providing "no guarantees that the giant facility will yield more discoveries than current or planned facilities." Dr. Arno Penzias, a Nobel laureate in physics at the Bell laboratories, was quoted as a leading critic worried about the sacrifices that would be made in order to facilitate the completion of this project.

The SSC project, as it was called by 1987, rivaled another fantastic project undertaken by the Reagan Administration, the so-called Star Wars (SDI) project. In February 1987, the *New York Times* reported that $5 billion had already been spent on the antimissile program, and that another $5.9 billion was being requested for that fiscal year (Broad 1987). Mind you, though this was a top-secret project established to fight the evil empire, the Soviet Union, and world communism, it was unclear how effective, if at all, this expensive project would be. From this perspective, what was a mere $4 to $6 billion if the SSC project could ensure the glory of science and the legacy of a presidency? Recall John Locke's admiration for the monuments of the master builders of his time (i.e., Newton), and imagine the desire to facilitate the creation and establishment of new monuments that would enhance an American reputation around the globe. What the British empire was for the nineteenth century, the United States could finally become at the end of the twentieth century, and this stature would not be limited to the horrible atomic bomb in World War II, but would be accomplished in the name of pure, basic science!

By April 1987 there were enough scientists involved in the project to call into question its entire design and the material used for conductivity. Some suggested the possibility that new magnets might allow the tunnel to be only five as opposed to fifty miles long. At issue was not only the timetable and costs projected by the administration, but the sense that by the time the SSC was constructed it would be obsolete (Gleick 1987). So-called pure, basic science was saddled with the significant concerns of engineers, so that technology was no longer portrayed (as in classical textbooks) as the handmaiden of science but as an integral partner or ingredient in the very conception of experimental designs. Hence the need to speak of technoscience.

But these concerns seem to have been overlooked by the time the Department of Energy announced on November 10, 1988 the site where the tunnel would be built. Out of thirty-five sites in twenty-five states originally proposed, and out of seven finalists, Texas was chosen. Senator Phil Gramm, a Republican from Texas, described the significance of the project in poetic language: "In high-energy physics, we, today, are basically looking at the fuzz on the peach; we know very little about the inside of the peach." (Franklin 1988) Because Gramm was a Republican Senator when Vice President George Bush (also of Texas) was elected President following a two-term presidency of Republican Ronald Reagan, it is hard to believe that politics did not influence the process of site selection.

It is cited that the State of Texas offered $1 billion in incentives to reduce the total cost of the project, a project that by this juncture had received only $200 million. Why would the voters and legislators agree to such an incentive? It was estimated that in addition to the original construction funding for the local labor market, there would be a federally funded budget of about $270 million annually. Simple arithmetic shows that the entire SSC project was an investment for the future and not a current expenditure in the state or federal budget.

By May 1990 two variables changed enough to cause some journalistic alarm, if not change the minds of politicians committed to the project. First, the project's cost estimate climbed to $8 billion from previous estimates of $4 to $6 billion, and second, Sigma Xi, the scientific honor society, conducted a national poll of 3,332 scientists and found that only 2% responded favorably to the construction of the SSC, as compared with 4% of the same sample group responding favorably to the expenditures related to Star Wars (Browne 1990). If even scientists did not speak in unison, how could the general public have been convinced that the most expensive project in basic or pure science was worthwhile?

Since they were not approved individually by congressional committees, other scientific projects undertaken by the Department of Defense were not under public scrutiny, so the prominence awarded to the SSC project balanced its potential results in light of its costs. While Star Wars had ideological appeal, the SSC was conceived as the new playground for scientific nerds. Moreover, Star Wars had the classic TV series *Star Trek* as a visual counterpart and explanatory device in popular culture. There was no visual or popular appeal to an underground tunnel in the middle of Texas. And finally, while Star Wars was construed as a military deterrent that could save the world, the SSC project had the potential to revive images of Los Alamos and the nightmares of another Hiroshima or Nagasaki.

Assuming that scientists are self-policing, as the ideals of scientific inquiry and the scientific community have been understood traditionally, it should come as no surprise that any scientific project, expensive or not, will have critics. If consensus among scientists cannot be expected (see Fuller 1988, Ch. 9), can there be some consensus among economists and politicians? By June 1990 it seemed that even the political arena found enough dissenters to argue that $8 billion was too much to spend for the SSC project. Therefore, one could either choose to drop the project or find sources outside the Federal budget to fund it. Congressional commitments and presidential aspirations remained bound to the State of Texas and to national pride, perhaps in that order. So the alternative chosen by the Bush Administration was to approach Japan as a possible investor-collaborator in the project. Japan was asked to invest no less than $2 billion. At this juncture the calculation was as follows: $5 billion from the United States government, $1 billion from Texas, and $2 billion from Japan and other countries (Sanger 1990). In October 1990, an unlikely candidate emerged as a potential $200 million contributor: the Soviet Union. The target of Star Wars just a few months earlier, the Soviet Union turned into a potential investor in the SSC after the fall of the Berlin Wall in 1989. In October 1991 the Japanese were asked to contribute $1 billion, and in exchange they demanded a bigger role in a number of other joint projects.

What used to be rhetorically a patriotic cry for scientific supremacy by the United States was toned down and relabeled in the popular press an "international project" by October 1991. The press suggested a Japanese "equity interest" in the project and estimated Japan's financial commitment to be between $1 and $2 billion. There was almost a tone of desperation in the media, for it became clear that the United States would not pay the full cost of the now $8.4-billion project and therefore that it might not be funded at all (Sanger 1991). A conciliatory voice from the Bush Administration made its rounds in the press, so that the very completion of the project, no matter under what conditions of compromise, was considered more important than the claims made as early as 1983 by the Reagan Administration. In January 1992, Japan gave no positive response to the ongoing request of President Bush, and the entire joint effort was contextualized in terms of the $41-billion trade deficit Japan inflicted on the United States. Was Japan not morally obligated to make up for this deficit with a little (less than 5%) contribution to the great glory of science? Had science lost its transnational, universal appeal?

By June of 1992 the House of Representatives voted to scrap the SSC project: that is, it eliminated $484 million earmarked for the SSC and appropriated the money to another bill. Five weeks later the Senate Appropriations Committee allocated $550 million to the project (Krauss 1992). (Political parties and lobbying interests can keep this game alive, playing the House off the Senate, so long as funding is sustained. This way everyone wins: there is enough money for economic stimulus to a scientific sub-economy, while there is not too much money to claim waste and fiscal irresponsibility.) In September of 1992, President Bush agreed to reconsider his opposition to a nuclear-test-ban bill if the SSC project could be saved. As reported, at stake were not only the economic benefits for Texas, but also Bush's concern for a political base for his re-election effort (Rosenthal 1992).

In an interesting development, by June of 1993, the media reported an escalation of SSC's costs to $11 billion and linked it to the $4-trillion national debt. It was also reported that scientific research had made up 5.2% of the Federal budget in 1978, but only 1.7% by 1993 (Krauss 1993). Big Science, so to speak, was losing its prominence and importance in contrast to Medicaid, Medicare, and Social Security. As economic conditions worsened or were perceived to have worsened, it was not surprising to see a front page byline: "House Kills the Supercollider, and Now it Might Stay Dead," intimating that previous attempts had been made but that the project had somehow managed to survive. It was not so much the House vote that killed the project, but the increasing perception that Senate support, so crucial for continued funding, could no longer be sustained. Most shockingly, it was reported that $2 billion had already been spent on the project by October 1993, while an additional $640 million was considered necessary to close the project down (Wines 1993). Recalling the figures mentioned earlier, it seems that by the time the project was buried in Congressional files, about $2.6 billion had been spent, more than the National Science Foundation spends annually for all scientific projects.

The SSC is not an economic project to be tossed across a political minefield. I have chosen to chronicle its history because it is an exemplar for postmodern technoscience. Though the context of my review has been the discourse of popular media rather than the reports of the learned technoscientific discourse, and though it would be presumptuous to generalize from this set of newspaper reports, I find it useful to underline some postmodern themes that emerge from this review.

First, it was not clear whether there was sufficient theoretical substantiation of the claims some physicists made concerning the potential for discovery that this project would have afforded. As such, then, this project was self-legitimating in the fullest sense of the term.

Second, there is an interdependence between the scientific and technical communities; without the engineering prowess to build the underground tunnel, no scientific feats could have been imagined. Moreover, the co-dependency is such that the failure of design by any member of the team will bring about the collapse of the entire project.

Third, there is a co-dependency between the technoscientific community and the political; without congressional support, this project could not have been accomplished. The political community is co-dependent on the economic; without cost projections, not only of the project installation but also of the infrastructure, there could have been no political decision. Moreover, without a cost-benefit analysis, one that appeals to the general public, there is no reason for political support.

Fourth, though decided on a federal level and with a national interest in mind, the specific political and economic interests of a local community that drive a national decision to approve a project and choose a site illustrate what may be meant by the notion of postmodern fragmentation. Of course, this statement must be contextualized in terms of postmodern capitalism, as David Harvey insists:

> The "otherness" and "regional resistances" that postmodernist politics emphasize can flourish in a particular place. But they are all too often subject to the power of capital over the co-ordination of universal fragmented space and the march of capitalism's global historical time that lies outside the purview of any particular one of them. (Harvey 1989, 239)

Fifth, national interests in the age of postmodern multi-nationalism are subservient to international pressures that cannot be measured in one dimension alone: trade deficits figure into the appeal for a technoscientific project. So, between localized regionalism and global instrumentalism, the old-fashioned nationalism of sovereign states is obliterated.

Sixth, the multitude of discourses employed in the debate over the continuance or discontinuance of the project illustrate that no discourse can remain insular in the postmodern age, nor does any single discourse have authority over all others. Whether radical or not, translation between various discourses is essential, for they are all vying for attention and power in similar fashions. Perhaps a more rigorous translation would have contextualized the whole decision-making process and allowed the project to survive. As it turned out, unforeseen variables and vocabularies took on a more prominent role in the debate and eventually brought about the demise of the project, or as some would argue, saved America from the embarrassment of a huge, useless tunnel in the midst of fire ants' hills.

Continued reports in the media tried to explain the demise of the project. Was it the new cost estimate of $13 billion, an untrustworthy figure that would probably increase again? Was it the lack of initial consultation with foreign governments? Was it domestic economic conditions? Was it the lack of consensus in the scientific community? But what emerged most prominently in articles published in 1994 were the broad implications concerning the scientific community and its conduct in the century to come. Is Big Science dead? Could it be replaced? Headlines such as "Big Project Dead, Physicists Find Small Is Beautiful" found their way into the public eye (Broad 1994), and as far as political reality was concerned, Big Science, at least the one on public display, was definitely dead.

By contrast, as far as Joseph Agassi is concerned, Big Science is neither dead nor should it be killed. Rather, what is dead is the particular bureaucratic form of operation it has become since World War II. Big Science must redirect its energies away from the hierarchical model espoused by Thomas Kuhn and adopt radical work reforms in order to ensure its future, whether as Big or as Smaller Science (Agassi 1988). The relation between the technoscientific model of Big Science and the postmodern political economy (as discussed in Chapter 6) is made apparent in this view of the future of Big Science.

Weimar Culture and Quantum Mechanics

What I find fascinating about the history of the SSC project is that one can retell its story without much concern for its theoretical underpinnings. This does not mean that the theoretical credibility accompanying its formulation is unimportant or negligible; rather, it may illustrate that economic conditions and political realities take precedence over the prestige and self-legitimation of science. Put differently, this is a case study of a situation in which the political leadership had sufficient doubts about the costs of a scientific experiment (though the SSC was never named so) to override the leadership of the technoscientific community. But even this characterization is oversimplified, for though it moves from technoscience-versus-politics to speak of leaderships and communities, it overlooks the subtleties associated with the factions of both the political and the technoscientific leaderships. For example, does it matter that the Texas delegation lost its power after President Bush's defeat for a second term? Does it matter that superconductivity is not a premier area of research in late twentieth-century physics? Does it matter that the costs could neither be contained nor predicted? How should we approach this whole situation?

In Chapter 6 I shift my general discussion from the inevitability of reality—in the present case a very amorphous reality whose particles appear and disappear in flashes of energy combustion—to the reality of the economic conditions under which technoscience is accomplished. I propose that there is a postmodern spin to political economy, for this provides a reconception of age-old models, whether those proposed by capitalists or socialists, as well as an appreciation of the interlocking of economic conditions with political aspirations. As the present case shows, whether one explores quarks or clouds, there is a price tag associated with any exploration.

Big Science today does not command the kind of public enthusiasm that the development of the atomic bomb commanded during World War II using the rhetoric of fighting fascism. America is not currently engaged in a war similar to the magnitude of a world war, one that threatens its very existence. The only war still preoccupying the public may be the one launched during Lyndon Johnson's administration, the war against poverty. But what does poverty mean, and can this country even afford to fight this war? Some would rephrase the question: can this country afford *not* to fight this war? And here an entire range of issues come to the fore, from the statistical measuring devices we use and abuse, to the ideological attachment or rejection of the homeless, to a comprehensive health care policy, and work rules for those on welfare. In this respect, what is at issue is not limited to the economic conditions under which a nation can support its "war effort," but its very Constitutional commitment to equality, liberty, and justice.

One may claim that this is another unnecessary digression, for how can the welfare system be relevant to technoscience policy? But if one appreciates the discussion of welfare or the general conditions of a society, one may appreciate a cultural milieu within which ideological convictions become so transparent that even the technoscientific community, the one that supposedly insists on "value neutrality and the consequent political immunity . . . [that] made it possible for scientists to survive under and gain support from regimes and groups representing the entire political spectrum" (Ben-David 1984, xx), cannot ignore them. Though exploring the Weimar culture and not our own, Paul Forman illustrates how the discourse surrounding quantum mechanics and causality was politically infused through and through. It may be informative to follow some of the details mentioned by Forman because they may illuminate the underlying presumptions associated with the debates about the SSC project, and with my concerns with pedagogy in Chapter 7.

Forman quotes Werner Heisenberg as having said in the spring of 1927 that "because all experiments are subject to the laws of quantum mechanics" and because "quantum mechanics establishes definitively the fact that the law of causality is not valid," it must be the case that the failure of causality is not limited to theories but to reality itself (Forman 1971, 105). What is at stake in upholding, as Einstein is reported to have done, the principle of causality during this period? At stake is not simply the applicability and neatness of a set of mathematical equations in a model proposed by theoretical physicists. Instead, the very definition of physics as a natural science, the very definition of the natural sciences in contra-distinction to the social sciences, and the very scientific apparatus of human knowledge are in jeopardy. If causality is no longer the bedrock of knowledge, if there is no explanatory or predictive model on which to base life's decisions, then the entire cultural orientation of the age must change as well.

Causality's definition during this period (1918–1927) shifts from "the postulate of the lawfulness of natural processes" to "rigorous determinism" (Ibid., 69). The connection between causality and determinism is fastened in the mind of the public; there is a perception that causality bespeaks mechanism (as in the mechanical worldview of the scientific revolutions of the seventeenth and eighteenth centuries); mechanism is perceived both in deterministic terms and the terms of rationalism (at least in the sense of logical deductions); and there is a certain cultural expectation against which physicists construe their public pronouncements (Ibid., 101). In this sense, then, Forman reinforces the postmodern sense of the cultural context of the technoscientific discourse as it is mentioned below in Chapters 2 and 4.

But is the principle of causality, as it is understood publicly, applicable to quantum mechanics? Heisenberg says no. So does Erwin Schrödinger in 1921. As Forman says, Schrödinger maintains that the principle of causality is the postulate

> "that every natural process or event is absolutely and quantitatively determined at least through the totality of circumstances or physical conditions that accompany its appearance." But "in the past four or five decades physical research has demonstrated perfectly clearly that for at least the overwhelming majority of phenomena, the regularity and invariability of whose courses has led to the postulation of general causality, the common root of the observed rigorous lawfulness is—*chance*." Now insofar as the physical laws are statistical, they do not *require* that the individual molecular events be rigorously causally determined. (Ibid., 87)

Schrödinger sets up the seemingly contradictory postulate that within the rigor of causality in general one can find chance. Phenomena are determined now by probabilities, so that individual events are not reducible to a mechanical configuration of a direct and immediate cause and effect. Causality, as previously understood, is shown to be invalid in quantum mechanics, and this situation is not accompanied by regret and loss of faith in everything scientific, but is instead "greeted with relief and satisfaction." (Ibid., 105) Max Born, for instance, characterizes in 1928 the laws of physics under the spell of the deterministic definition of causality as having no place in them "for freedom of any sort, whether of the will or of a higher power," a view valued by good rationalists, but a view whose time has passed (Ibid., 107)

Niels Bohr pushed Born's claims to their logical edge in the same year by characterizing the quantum-mechanics postulate as containing inevitably the "feature of irrationality." (Ibid.) The infusion of irrationality into quantum mechanics goes far beyond Schrödinger's attempt to reconcile, though at different levels of analysis, chance and determinism or rigorous logical deductions with probabilistic experimental results. While Schrödinger shied away from introducing any irrationalism into the physical world, Bohr let the principle in through the front door. Why were some German physicists so eager to allow a dose of so-called Romantic irrationalism into their midst? As Forman concludes, they were delighted to turn quantum mechanics into an *acausal* theory "because of the irresistible opportunity it offered of improving their public image." (Ibid., 108)

So public image was relevant in the German Weimar culture, according to Forman, as it was important half a century earlier in the British empire (Sassower 1993a, Ch. 1). Notwithstanding the claims for political immunity that appeared after World War II, scientists clearly understood that as part of an intellectual elite, they were part of the culture. This was especially true in the case of Germany, where academics were and are public servants who depend on the largess of the government for their income and research funds. Would the SSC project have fared better if its supporters improved their public image? What ideological values did they neglect to highlight? Why did they fail to speak of political liberty and individual freedom in the face of the harsh realities of everyday life, as their German counterparts had done seventy years earlier?

Small Is Beautiful

While scientific research in America has become centralized and militarized since World War II (Remington 1988, 51–55), especially in light of the Strategic Defense Initiative (Star Wars), there has been a growing appreciation of the need for a "new social contract" between the scientific and the nonscientific communities (Ibid., 55–62) such that free inquiry, accountability to funding agencies as representatives of the people, and responsibility for applicability for the benefit of society as a whole, all become evident to the public. Whether the general funding of basic research has increased or decreased relative to national expenditures during and after World War II is not the issue; rather, it is whether the expenditures fall exclusively under the guise of national security and secrecy. Does the public deserve to know what is done with its money? Should the public participate in decisions that will affect its well-being not only for the short term but for many years to come? As we have seen with the development of nuclear energy, far beyond the construction of power plants there is the problem of disposing of nuclear waste in a manner that has the least secondary, polluting effects on the environment, whether it is in the deserts of the wild west or in poor neighborhoods where dumps are oversaturated.

Within this social context, it was reasonable for the scientific community to propose the SSC project, a project of basic science whose military applications remained unmentioned in most of the literature. In 1985 there were two hearings before the Task Force on Science Policy of the Committee on Science and Technology, a standing Committee of the House of Representatives. In these hearings it was first argued that there was a "natural time scale" for such projects as the SSC, and that the time scale was "now" (Congress 1985a, 79). It is fascinating to review the urgency with which funding was requested: with the assumption that the technology was readily available and that if its use were delayed, the window of opportunity would diminish or be completely closed. Under these conditions, there was also a call for international support and a funding formula that would include international collaboration (Ibid., 170).

The second set of hearings took place in June, and the key expert witness was none other than Victor Weisskopf, whose scientific credentials were touted before the committee like a prize bull at a county fair. Weisskopf provided a long-term perspective, looking back some forty years and then forward to the next century. His overview, therefore, was comprehensive enough to attract the attention of committee members and the entire scientific community. It was perceived that he was not speaking with the self-interest accompanying zealots, but with the maturity of a scientific guru or an old sage whose life experience is valued. He addressed three key issues in his testimony: one pedagogical, a second economic, and a third about the nature of scientific research.

First, Weisskopf emphasized the future of science and the enticement that such a big project would have for young people to join the efforts of the scientific community. As far as he was concerned, the project should be large enough and far-reaching enough to enliven the imagination of a young generation of scientists so that they recognize the opportunities open to them for future research. Instead of encountering a scientific apparatus to which they can contribute nothing, they should perceive uncharted territories that beg their involvement and long-term commitment.

Second, Weisskopf reviewed all the known installations of high-energy accelerators around the world and explained that the European centers and national budgets supporting them are overextended. To expect, as the first set of hearings before the Committee suggested, an international collaboration mandating European contributions toward the SSC project seemed, therefore, unreasonable if not outright impossible. So it would make sense for the United States to undertake the development of a new generation of equipment with which to explore particle physics by itself. If this option seemed unfeasible, given the economic conditions of the Federal budget, then perhaps the idea of "user fees" could be explored. As it stands, Weisskopf said, scientists are expected to help their international counterparts in research by allowing them to use their "machines" on request. This puts a tremendous burden on existing centers and their dwindling budgets. If every center had basic national support, it could also be used by others who paid for its use, and it could serve more scientists without having to rely exclusively on the hosting country's good will and generosity. (This idea seems to follow Milton Friedman's view of user fees for everything in the public domain, from highways to national parks [Friedman 1982], and can be reconceptualized to fit the postmodern orientation suggested in the next chapter.)

Third, Weisskopf reminded the committee of the 1950s, when Enrico Fermi requested funds for an accelerator in Chicago (now named after him). Did he have to explain what it was good for, or how much it would cost? No! It was understood that if he requested it, he should be obliged; results would follow as surely as day follows night. Was this a mistake? Was it necessary to have hearings before Congress to grant him his wish?

You see, argued Weisskopf, we scientists are trustworthy and we eventually "deliver" the goods you expect of us. It may take some time, it may take some money, and it may take the energy of great minds, but no matter what the expense, the fruits are there to be had, almost guaranteed (Congress 1985b, 1–43).

Weisskopf's testimony overshadowed any concerns and doubts that lingered among committee members. As we have seen from the reports in the popular media, funding was granted (to the tune of $2.64 billion) and our representatives felt comfortable supporting the SSC project. Yet it is interesting that the rhetoric of the *New York Times* differs from and does not account for the rhetoric of Congressional hearings, where policy decisions are eventually made on behalf of the public. Why did Congress vote for the project initially? Perhaps the answer can be understood in this way: first, the SSC did not have the negative connotations of the results of the Manhattan project; second, it did not carry the far-fetched promises associated with the SDI project; third, it was about basic science, true knowledge, as wholesome as apple pie. But there was one caveat: it was impossible to undertake this effort piecemeal, so it had to be done in a big way if it was going to yield results at all.

Instead of pursuing an analysis of communication techniques and America's political structure, I shall move on to examine briefly some competing views in developmental economics, because they are concerned with projects similar in magnitude to those termed Big Science.There are two basic models of development, one trumpeted by Schumacher, the other by Rosenstein-Rodan. Schumacher urges us to think of local communities in which development is to take place, and about the humans who are the recipients of development efforts. Because of his concern for people and the environment, he pleads for what he terms intermediate technology, or what others have called appropriate technology—that is, a technological application custom-tailored for the context in which it will be applied. Instead of bringing heavy equipment to India for housing developments, for example, it may have been more beneficial to create smaller machines that use less fuel, are operated by more people (since the labor market is in dire need of work and is cheap), and do less damage to the environment (Schumacher 1973).

By contrast, Rosenstein-Rodan argues that piecemeal reforms and small steps toward development are ineffective, even inappropriate. In order to develop a region or a country, it is necessary to develop its infrastructure as much as the factories or equipment brought in from outside. What use is there for a factory that produces shoes if there is no road on which to transport raw materials or finished goods? Even if a road is built, where does it lead? Is it not necessary to have shipping ports, ships, and airplanes? Unlike Schumacher, who holds that "small is beautiful," Rosenstein-Rodan argues for the "big push" view of development (Rosenstein-Rodan 1958).

From what we have seen above, it sounds as if particle physics must follow Rosenstein-Rodan's view, because Schumacher's view is inapplicable. But is this the case? Must one construct a fifty-two-mile tunnel to discover the workings of quarks? Can this not be simulated on a computer? Even if simulation is insufficient, are there no other alternatives? As it turns out, there are. In April of 1994, there were reports that several attempts had been made to experiment with high-energy physical entities under conditions different from those presumed to be required in the SSC project. (The 1929 invention of the hand-held cyclotron by Ernest O. Lawrence was cited.) A team at the University of California in Los Angeles had promised to deliver the same results envisioned by the SSC team. At the Argonne National Laboratory in Illinois another team was developing a different technique to rival that of the UCLA team. And a new acceleration method was being tested at the Brookhaven National Laboratory on Long Island (Broad 1994).

Will this barrage of experiments yield interesting results? Definitely, yes. Will all of these experiments cost as much as the SSC project? No, their total costs are but a fraction of the already-spent $2.64 billion. So should society endorse the slogan "small is beautiful" in the context of particle physics or high-energy science?

Postmodern Technoscience

There is a need to conceive of the future in terms of postmodern political economy, a way to orient decision-making processes on national and international levels (to be explored in Chapter 6). With this orientation, there is a definite need for centralized information flows, and decentralized operational and intellectual flows. That is, while it is crucial to plan the future of a region or a nation and therefore necessary to share information with a wide variety of people and institutions, it is also crucial to provide conditions of individualized operation with the elasticity now possible in the computer age. Perhaps it is worth revisiting contemporary commitments to copyright and intellectual property in the terms proposed by Collingwood (1958, Ch. 15) in the context of the production and consumption of art: what is wrong with copying a masterpiece? What is wrong with copying experiments whose details are shared in the literature or at professional meetings?

In the case of accelerators, it is important to share information among all scientists working in the field and among those not directly involved in the field. First, so that duplication will be maintained at the minimum required for checking previously achieved results. Second, so that economies of scale will allow minimum waste of resources: why reinvent, rather than improve, the wheel? Third, so that dead ends are made public and not reached repeatedly. Fourth, so that alternative techniques are developed where dead ends are known. Fifth, so that equipment is used more efficiently by teams at different centers of research. And sixth, as Aharon Kantorovich argues well from historic case studies, expertise and experience from one field can sometimes be transported into another field with amazing results (Kantorovich 1993).

I take it for granted that scientific results, like the results from any other activity, cannot be guaranteed in advance, and that there is a certain level of gambling in all funding for research and development. Why not, then, spread the risk the way prudent gamblers do? Why not underwrite the operations of many more centers of research and universities? My appeal here is not the naive appeal of those wishing to go back to the so-called good old days of Galileo and Newton. Instead, it is an appeal that accounts for the economic reality of the current state of affairs, one that would switch us from poker tables to the tables of coffee shops.

I do not wish to portray a rosy picture of the intellectual world, for it is rife with competition and jealousy, rivalry and prestige, back-biting and outright fraud. Yet there is room for change under the conditions of postmodern political economy. We are bound to consider political realities in the development and control of technoscience; we are bound to consider economic conditions that permit or limit the research and development in technoscience; and we are bound to consider the multiplicity of perspectives and goals associated with technoscientific projects. In some ways we are just like Molière's protagonist, who did not know that he had always been speaking prose.

Assume, then, that we are indeed looking into a twenty-first century of postmodern technoscience. How do we prepare for it? What conditions must we change to accommodate this inevitability? What circumstances can we change? What power do we really have to engage in such an activity? With these questions in mind, I turn to the rest of the book. Some chapters examine the political constellations and economic circumstances under which postmodern technoscience can accomplish those tasks that will benefit ourselves and our environment. I end the book with a chapter in which I examine art and imagination as means for a pedagogical transformation that will enhance human creativity and satisfaction in the next century.

But before I get there, I try to explore my role as translator so that I can appeal from more than one perspective to more than one goal. The failure of translation may result not only in the failure of projects, like the SSC, but also in a failure to converse at all. The failure to converse may result in a failure to see why cultural influences and different power relations (of technoscientists, politicians, academics, etc.) may need to change under changing demographic and economic conditions.

Chapter 2

Postmodernism and Technoscience

Prologue

This chapter is a journey that begins with Karl Popper, joins Paul Feyerabend for a while, and then seeks the company of Donna Haraway. Subsequent chapters will engage additional companions; I leave their introductions until their later appearances in the text. All of these companions contribute to a postmodern orientation, perspective, or sensitivity as it applies to science. The sensitivity I have in mind is that of concretizing epistemological debates and suggesting the need for a psychological setting different from that of the positivists and for a politics different from that of the elitists. The application of this sensitivity might lead to a postmodern technoscience, even when the specific concerns of the leading postmodern writers are different from those of philosophers of science and vice versa. The two parties ignore each other.

I engage them both here, hoping to acquaint them with each other, well in accord with their respective theories though regrettably not yet exemplified in their practice. My brief sketch is intended to introduce critically postmodern themes, perhaps tease some issues that characterize postmodernism. Those interested in technoscience may find it helpful to learn this vocabulary or at least the concerns that inform it, however obscure it all may seem from a distance. As such, then, the present sketch is incomplete.

For example, Popper is the great developer of the last major attempt to demarcate science from all other forms of discourse, and Feyerabend criticizes this attempt. Now, the "problem of demarcation" (no matter how it is narrated) is in fact a political problem, one that bespeaks a form of privileging—a "metanarrative" (Lyotard 1984, xxiii–xxv)—that the Lyotardian "postmodern condition" refuses to accept in principle. Popper's demarcation forms a certain boundary between "fiction" and "science," and that boundary translates into the fence of an exclusive club, a status symbol of positions of power, and as such, it is an obstacle on the road of critical analysis of the scientific culture in which we live (Ormiston & Sassower 1989, Ch. 2). The demarcation itself creates a situation that overlooks some of the very issues raised by Popper himself.

Popper claims that metaphysics does matter in science, that Truth must be relegated to the position of an honorary concept in the scientific community but one without any political clout (much like the Queen of England), and that the unresolved problem of philosophy of science remains psychology: how do scientists come up with new ideas? What disposition must one have to be able to challenge established models and propose alternative ones? But still, how does psychology fit into the context of scientific discourse?

The answer can be offered with the aid of tools from the discourses of postmodernists. In some respects I find solace in the psychological possibilities that open up within the critical discourses of postmodernity, in the sense of empowering all narratives and reorganizing institutional power relations (Sassower 1993a, Chs. 5–6). Of course, one could say that Michael Polanyi (1958) and Thomas Kuhn (1970) have given us as much as we need to know about the workings of what they refer to as the "scientific community," analyzing issues of leadership and responsibility, raising questions about the ways in which the methodology of science can and should be pushed further along the paths of a research program, with or without the tampering that can be had with its metaphysical "hard core" (Lakatos 1970).

For my taste that is not enough. I recommend that we radicalize the critiques of science in the wake of postmodern literature, which, regrettably, overlooks too many aspects of the scientific enterprise. This radicalization would deal with the political character that has been detected in the domain classically renowned for some alleged value-neutrality. As already illustrated in the last century, when Karl Marx attacked classical political economists, he criticized the theoretical content of their models as well as their political conclusions, not to mention their methodological foundations. As such, he invoked the connection between "theory" and "practice."

Contemporary feminists continue their critiques of the sciences in a similar fashion, locating scientific research within its socioeconomic and political context. I end my discussion in this chapter with Haraway, whose socialist-feminist views on science are essential for the appreciation of what should count as postmodern technoscience. Its variations and differences are contrasted, for example, with some pseudo-liberal ideals expressed by Richard Rorty.

Background

"Postmodernity" has suffered suspicion and ridicule since its various incarnations in literary and philosophical circles. Perhaps most noticeable has been the refusal to grant postmodernism any status whatsoever, that is, the refusal to acknowledge its very existence as a mode of thinking and doing, if not as an intellectual movement. For example, C. Barry Chabot claims:

> (1) that no satisfactory and widely accepted account of postmodernism now exists; (2) that much of what is called postmodern in fact derives directly from modernism; and (3) that most arguments for its existence achieve their initial plausibility largely through improvised characterizations of modernism, especially characterizations that neglect its nature as a second-order concept. (Hoesterey 1991, 37)

Others, like Martin Jay, have warned against "a leap into the postmodernist dark" (Ibid., 108), as if there is something dangerous about the psychological settings and possibilities offered by postmodernism or as if postmodernism would return us to a dark age.

One way of understanding the psychological settings of postmodernism is to emphasize that broadening the horizon of possibilities—of narrative production and personal exchange—does not entail forming hierarchies of choices. As Matei Calinescu says: "What aesthetic dualism or pluralism shows is simply that a choice does not necessarily imply a summary dismissal or ignorance of other available alternatives." (Ibid., 167) There is a difference between the openness of postmodernism and the openness of objectivity and rationality as the guiding principles of modernity. The openness of postmodernity attempts to blur disciplinary boundaries and encourages the violation of traditional rules; it suggests experimental techniques that juxtapose theories and practices usually kept at arm's length from each other because they are either incommensurable or historically differentiated (Sassower 1993a, Ch. 4).

The reception of postmodernism illustrates that the articulation of different forms of discourse does not itself necessarily enhance the possibility of greater engagement across intellectual barriers, an engagement whose political potential has been sought all along by its proponents. At the same time, "science"—that category of modernity that has supposedly fought the prejudices of church authorities and the superstitions of the uneducated—has lost its leverage as "peacemaker." It is perceived to be a category of power (white, male, hegemonic, etc.) and domination/oppression that eschews debates when these debates are deemed too critical (and therefore threatening to its authority) or not critical enough (and therefore not worthy of attention at all). In attempting to engage postmodernism with science, I try to bring together the potential each offers separately in order to have a greater impact on critical discourse.

I will return to the question of critical and radical engagement of postmodernism and science after briefly examining what may be considered the tenets of postmodern technoscience, clarifying along the way why I find it important to focus on technoscience and its attendant critiques. In the meantime, perhaps David Harvey's advice can suffice:

> In a sense it does not matter whether postmodernism is or is not on its way out, since much can be learned from a historical enquiry into the roots of what has been a quite unsettling phase in economic, political, and cultural development. (Harvey 1989, ix)

Harvey's advice concerning postmodernism can be extended to the notion of technoscience, the coinage of which may remain a contested issue of authorship. In 1982, Jean-François Lyotard writes that:

> In the present epoch, science and technology combine to form contemporary technoscience. In technoscience, technology plays the role of furnishing the proof of scientific arguments: it allows one to say of a scientific utterance that claims to be true, "here is a case of it." The result of this is a profound transformation in the nature of knowledge. Truth is subjected to more and more sophisticated means of "falsifying" scientific utterances. (Lyotard 1993, 14–15)

Of course, the combination of the terms science and technology into technoscience is still explained here in traditional terms—technology is the implementation of science, a form of testing scientific knowledge—and not as a blurred site of knowledge production, that is, a site wherein one cannot do without the other (Ormiston & Sassower 1989).

Yet Lyotard inadvertently brings the Popperian perspective into prominence by comparison to some other positivist moves on behalf of the conjunction of science and technology. Whether traditional or not, positivist or falsificationist, the term technoscience is useful as an acknowledgement of the close relation between seemingly different modes of knowledge production and dissemination. In a way, the term alerts readers and practitioners to new needs and synergies whose time has definitely come by the late twentieth century.

Should There Be Postmodern Philosophy of Science?

The question, Should there be postmodern philosophy of science? addresses two related contentions: (1) postmodernism does not differ from modernism (as quoted above), and (2) if there is a difference, then there has always been a postmodern philosophy of science. Indeed, if one were to agree that postmodernism is in general characterized by its psychological settings and possibilities, then it is possible to illustrate that postmodernist attitudes, such as a recognition of the interpretive activity required to "explain" or write about "nature," have been observed in many bona fide scientists, such as Max Planck in the twentieth century (Greenberg, 1990). Going back to the seventeenth century, Francis Bacon's sensitivity to the interpretive mode of scientific research is easily detectable in his *New Organon* (Ormiston & Sassower 1989, 27–29) and as such he can be viewed as an "adherent" to postmodernism. Rorty, for instance, would have to agree with this contention since he too sees Bacon as closely aligned with Feyerabend as a "prophet of self-assertion," as opposed to aligning him with the prophets of "self-grounding," Descartes and Kant (Hoesterey 1991, 92).

Methodologically, Rorty's comment on Bacon can be understood as advocating a view of philosophy of science that focuses primarily on questions of orientation toward the material under study. According to this view, epistemological questions are more directly connected to psychological dispositions and the fashions one adopts both historically and socially in terms of the environment in which the research is undertaken. So whether one agrees that a postmodernist flavor has accompanied even the most rigid modernist versions of scientific inquiry or that it is a relatively new or different kind of classification of and within philosophy of science, it seems that the examination of postmodern philosophy of science has to do with a constructivist view of the scientific enterprise. The construction of science, what Stephen Toulmin describes as the shift from "scientist-as-spectator" to "participant" (Toulmin 1985, 29), becomes at once a politicized undertaking, one that can no longer shield itself from public engagement because of a presumed value-neutrality.

Despite a general cavalier use of the term "postmodern philosophy of science," one would be hard pressed to find a clearly defined area of research that fits this term or a consensus among its advocates. Currently, several areas of study come under the label of philosophy of science: history of science, methodology of science, sociology of science, anthropology of science, even science, technology, and society. But the category of postmodern philosophy of science—combining two disputed discursive practices, postmodernism and philosophy of science—is at once intriguing and suspect.

When I think of fabricated categories, such as the sociology of science, I think of convenient ways of lumping together a variety of works by a diverse group of people. In doing so, I realize, we do injustice to both the works and their authors. Some authors object to certain classifications but demand others. For example, Bruno Latour claims to have coined the phrase *anthropology of science* "clumsily a decade or two ago" (Latour 1990, 146). According to him:

> We have never left the old anthropological womb—we are still in the old dark ages or, if we prefer, we are still in the infancy of the world. How will we call this retrospective discovery that we have never been modern? Postmodern? No since this would imply a belief that we have been what we have never been. I propose to call it *amodern*. (Ibid., 164)

Though Latour suggests a renaming of his and others' work as anthropological, his historically informed approach to the practices of societies does not reduce itself to what he terms the Durkheim-Mauss thesis that "there is some correspondence between the way we organize our society and the way we organize our cosmological classifications." (Ibid., 165) Rejecting Nature and Society as starting points, he suggests that "Nature and Society are now accounted for as the historical consequences of the movement of collective things." (Ibid., 170)

Whether he likes to admit it or not, Latour's anthropology is postmodern through and through; that is, it is informed by certain themes that permeate the postmodern vocabulary, such as the dismissal of transcendences, the ambiguity of language and the fluidity of words and terms, the multiplicity of interpretations that accompany the historical study of any phenomenon, and the breakdown of all hierarchical settings, be they metaphysical or political. Latour's conventions are borrowed. This does not mean that his insights are flawed or that his anthropological work in the laboratory (Latour & Woolgar 1986) is useless. On the contrary, Latour's work is significant because it crosses disciplinary boundaries without regret or apology, and in this crossing it critically evaluates any terminological and theoretical presupposition it finds either implicit in certain practices or explicit in certain theoretical models.

Put differently, works such as Latour's that defy traditional categories are of interest because of their balancing acts among a variety of methods and techniques of inquiry, all of which are used and discarded at will and without regret. The appeal, if any, is an appeal to oneself to be honest and forthright, to admit one's failings and declare one's accomplishments without, once again, arrogance or apology. Having said this, I should mention that this sort of orientation is culturally informed and informs the culture in which it flourishes. What makes "science" such a revered area of our cultural life, according to Robert Merton, are four "institutional imperatives—universalism, communism, disinterestedness, and organized scepticism" (Merton 1968, 607). Regardless of how far we have come from believing that these institutional imperatives are in fact verifiable in our culture—having encountered a variety of critiques and scandals of the scientific community—there is some lingering attraction to the promise science may offer.

What does the discursive labyrinth called postmodern science or postmodern technoscience in fact stand for? Is it a chronological or historical designation that demarcates between modern and *post*modern philosophy of science? Does it herald a "new" way of thinking about and doing science? Besides, is the focus science, technoscience, or the philosophy of science? As the rest of this book unfolds, I will use more readily the term "postmodern technoscience" as the term that most usefully contains and represents the issues under discussion here, and perhaps as a term that signifies all the questions that cannot be answered in a canonical, modernist vein. In the meantime, I wish to record briefly some protests and concerns voiced by contemporary critics who are still reluctant to accord any intellectual status whatsoever to the designation of an area of study associated with anything postmodern.

For example, Zuzana Parusnikova is frustrated by the possibility of a postmodern philosophy of science. She concludes that "the prospects for a postmodern philosophy of science are not particularly promising: a postmodern philosophy of science instead dissolves into various sociological, historical or literary studies of science" (Parusnikova 1992, 22). In the course of her analysis, Parusnikova laments the loss of the scientificity of philosophy and the foundationalist aspiration that made it a valuable endeavor. She echoes Latour's claim that postmodernism is "the most sterile and boring intellectual movement ever to emerge." (Latour 1990, 147) As far as Latour is concerned, postmodern philosophers "are simply disappointed by the whole Critique enterprise and fail to believe anymore in the joint promises of rationalism and socialism. They have not moved an inch beyond. In spite of their presumption this shows they are modern to the core." (Ibid., 169)

These discrediting moves are not new and are wonderfully traced in Ingeborg Hoesterey's anthology *Zeitgeist in Babel* (1991). Either postmodernism is a sham—that is, modernism with a different vocabulary—or it is such an enterprise that to associate the seriousness of the scientific enterprise with it is tantamount to blasphemy. There have been other ingenious ways of appropriating the terms *postmodern science* and *postmodern philosophy of science* in ways that make one wonder what reading lists circulate in the academy, by whom, and for what purposes. For example, Joseph Rouse associates the term postmodernism with Arthur Fine's view of science (and also Dudley Shapere's) that amounts to something like a "respect for the local context of scientific inquiry and resistance to any global interpretation of science which would constrain local inquiry." (Rouse 1991, 613) Nancey Murphy, for her part, associates postmodernism with "holism in epistemology and the theory of meaning as used in philosophy of language" (Murphy 1990, 291) so that the preeminent thinkers in her context are Quine and Wittgenstein.

Given the variation in the use of the term postmodernism, I should delight in any of their appropriations, because they illustrate how useful the term or category has proven to be. Yes, they may miss some important French ingredients in their baking, but their bake sale may prove successful enough to bring the audiences back for more—and perhaps the next time around the customers will buy some cakes that have a French flavor of the Lyotard or Latour varieties.

Having seen that a postmodern orientation can be traced to a number of historical texts, we may nevertheless detect a difference between so-called modernist and postmodernist versions of philosophy of science. At the same time, the novelty of the approach is not in terms of "originality," but rather in terms of difference. The reflexive conception of postmodernists, that their efforts are different, is a direct critique of previous or other attempts, such as the Enlightenment modernists who claim superiority (of approach, methodology, and success) over previous practices (relegated at that point in time to nothing more than superstition and wishful, unscientific thinking). The jargon of difference—whether in Jacques Derrida's (1988) or Lyotard's sense (1988a)—tries to ameliorate previous contentions of superiority by opening fields of research, offering additional options that in their introduction do not claim either superiority or the necessity to do away with all previously known forms of discourse. This is what is at stake when speaking of *displacement* as opposed to *replacement* (Ormiston & Sassower 1989). Moreover, the jargon of difference ensures, as Toulmin concedes, that it is no longer necessary to attempt a "positivist" reduction of scientific research to a "single set of methods." (Toulmin 1985, 29)

But so far I have said little about what postmodern philosophy of science *is*. Perhaps it is wise to take Susan Rubin Suleiman's advice and problematize this very question: "If postmodernist practice in the arts has provoked controversy and debate, it is because of what it 'does' (or does not do), not because of what it 'is'." (Hoesterey 1991, 113) Having changed the question from what it is to what it does, Suleiman continues to explain that postmodernist "doing" is invoked "in a particular place, for a particular public . . . at a particular time." (Ibid., 119) The contextualization of postmodern practices, as Charles Jencks explains, differentiates them from others: "Modernists and late-modernists tend to emphasize technical and economic solutions to problems, whereas postmodernists tend to emphasize contextual and cultural additions to their inventions." (Ibid., 9) But even this statement, as seen in the previous chapter in the case of the SSC, is itself contestable, for it is possible if not necessary to combine a political and economic contextualization with a cultural one.

Coming up with a more precise definition is quite difficult, because definitions tend to be too broad or too narrow, and then they need to be qualified forever, as Kuhn still has to do with his notion of paradigm. Moreover, providing a definition goes against the postmodern grain because it pretends to capture a moment that is too fleeting to catch, and therefore, some speak of "postmodernities" (Connor 1989, Ch. 2). Instead of definition, then, let me follow Suleiman's lead and add that postmodernist practices—those of philosophy of science included—belie and illustrate an orientation, an approach, an attitude, as has been my approach to expertise (Sassower 1993a). Whose attitude?

Being playful, Umberto Eco claims in an interview with Stefano Ross that one defines the postmodern nowadays as "everything the speaker approves of"; but he also acknowledges that postmodernism is "a spiritual category . . . a world view," one that is characterized by "irony" and a "metalinguistic play" (Hoesterey 1991, 242–243). Closer to the terminology I have used so far, Eco suggests that he "would consider postmodern the orientation of anyone who has learned the lesson of Foucault, i.e., that power is not something unitary that exists outside of us." (Ibid., 244) Perhaps it is important to note in this context that Eco wishes, along postmodernist lines of orientation, to insist that "our behavior in the world ought to be not *rational* but *reasonable* . . ." (Ibid., 244)—transcribing the epistemological discourse of science and technology into that of psychology and sociology and vice versa. Even more is at stake with this orientation: a lowering of the threshold of rationality to that of reasonableness, one whose elasticity is presupposed because of its amorphous definition.

Back to the specific orientation I wish to ascribe to or express through postmodern philosophy of science. What I have in mind here is an orientation that lies somewhere between Feyerabend's anarchism of "anything goes" and Popper's rationalism of falsification. Incidentally, my appeal to these thinkers differs from the appeal that Gerard Radnitzky, for instance, makes to Popper and Wittgenstein as the two polar opposites that "carve" the landscape of twentieth-century philosophy of science (Radnitzky 1991). For my purposes, as indicated above, the company of Popper betrays my own intellectual genealogy and does not set limits to what can and should be considered under the ambiguous title of postmodern philosophy of science or postmodern technoscience.

In its provocation to proliferate a multiplicity of scientific models and methods of inquiry, postmodernism adheres to a kind of pluralism envisioned by Feyerabend, and in a different manner also by Agassi (1991). In both cases, there is a flexible attitude toward the latest theory and scientific gimmick: they are welcome newcomers as long as the "old guard" is not dismissed or demolished. This is the critical engagement I have in mind, perceived as a process of displacement—where the "new" and the "old" continue to confront each other—as opposed to a process of replacement—where the newcomers do away with any traces of the old guard.

The issue is as much about the attitude displayed by each group toward the other, as each group fights over restructuring power relations. There is even some affinity on the question of relativism in both Feyerabend and postmodern conceptions: the criteria of evaluating models depend on a theoretical framing that needs itself to be contextualized culturally. Feyerabend pushes methodological debates to their logical extremes and cracks them open by inviting any method of inquiry to participate and be accorded similar (scientific) legitimacy to that of already established methods. Acknowledging Feyerabend's logical indebtedness to Popper's views, and with a tint of irony, I ask: what can Popper add to this openness?

Popper adds restraint to Feyerabend's presumed recklessness. The Popperian restraint does not limit potential "newcomers," for Popper solicits any and all hypotheses to be considered scientific as long as they refrain from shielding themselves from critical empirical testing. That is, no matter how wild a speculation (conjecture) is, it should be permitted to confront the scientific court of appeals and have its day in court. If it is falsifiable in principle, then it retains its putative scientific status until it is in fact falsified and a revision is proposed (see Popper 1959 and 1963). The Popperian restraint, one that can also be found in Feyerabend regardless of his sometime protestations to the contrary, maintains that critical examinations be based on both rationality and empirical data.

Is such a restraint too harsh? Will it rob postmodern philosophy of science of its French mystique and potential? The answers to these questions depend on the application of the restraint. Enlightenment adherents of the scientific method go to extremes to ensure a strict application of reason and rationality and to codify their applications in strict logical terms. By the same token (and with a great deal of overgeneralization), empiricists appealed to the empirical data in a rigid and faithful manner that overlooked some subtleties, such as the appreciation of the theory-ladenness of observations in Norwood Hanson's sense (Hanson 1958). According to Toulmin, what "went wrong" in the modernist project as it progressed into the present is not realizing that "the planetary system is a quite exceptional system. No other system of entities in the natural world lends itself to prediction in the same way." (Toulmin 1985, 30) There is a way to be rational without being narrowly committed to a specific set of methods or logical principles, just as there is a way of being empirical without becoming an empiricist. A balancing act of this sort may have informed Umberto Eco in setting the preference for being reasonable rather than rational. The rules that guide the reasonable come from specific contexts, while the rational appeals to transcendent rules.

The appeal of postmodern technoscience lies in its elasticity, its flexible adoption of the ideas proposed by Popper and Feyerabend together as well as those proposed from literary criticism, architecture, and art criticism. I will continue to focus at present on Popper and Feyerabend, for the juxtaposition of their ideas may show a discursive intersection attributable to postmodern philosophy of science, one that is echoed in feminist critiques of science. Feyerabend's principle is considered irrational, for he provides no strict criteria of demarcation between science and pseudo science, thereby legitimating *ipso facto* any theory and model (like Chinese as opposed to Western medicine) that works under certain conditions as scientific (Feyerabend 1975, Ch. 18). By contrast, Popper's principles are considered rational and his criteria of demarcation are considered strong (positivist?) contenders under the logical positivist standards of this century. So how does postmodern technoscience wedge itself between rational and irrational principles? Can it disregard this binary and employ both, and if so, at what price? That is, is the postmodernist elasticity not a vice rather than a virtue, one that allows contradictions to coexist?

It seems that postmodern elasticity can be tested on several levels of technoscientific investigation and practice, such as the adherence to a foundation, to the rules of logic, and to critical evaluation. First, because postmodernism refuses to accept any principles whatsoever as setting a permanent foundation for anything, the use of principles deemed sometimes rational and at other times irrational is unproblematic. That is, contextualizing situations does not mean relinquishing foundations altogether, but rather upholding them only within specific confines for particular purposes. A foundation whose appeal is situated is bound by specific "pragmatic" goals in the light of which it is applied (Ormiston & Sassower 1989, 17, 19). More on this in Chapter 3.

Second, even the anarchistic principle associated with Feyerabend is not presented in an irrational form; on the contrary, there is a long and arduous rational narrative that argues for the acceptance of this viewpoint and appeals to notions of consistency and non-contradiction. In this sense, then, one can appreciate a minimalist use of logical principles, no longer as strict rules of practice but rather as guidelines that enable the discourse to get off the ground, and perhaps even turn out to use logical rules against the use of logical rules. Once again, the attitude here is not one that disallows certain vocabularies or forms of discourse nor one that requires à la Jürgen Habermas a "consensus;" instead it is a liberating attitude, one that attempts to explain away traditional binaries by setting a low threshold over which anyone can pass (Agassi & Jarvie 1987).

Third, once critical evaluation is viewed as a tool for the improvement of general discourse, whether it is labelled scientific or not, the quest for scientific legitimacy on grounds other than critical evaluation becomes superfluous, if not mere appeal to authority. Critical evaluation, as the most simple possible threshold of communication, turns out to be the reference point according to which the rules of a critique are formed. Lowering standards is objectionable; as Clement Greenberg says: "Modernism, insofar as it consists in the upholding of the highest standards, survives—survives in the face of this new rationalization for the lowering of standards." (Hoesterey 1991, 49)

The elasticity of postmodern technoscience continues to occasion numerous criticisms and objections, some of which have already been mentioned. Reiterating several concerns at this juncture may help us to appreciate the issues raised by some feminist critiques. One objection to the elasticity portrayed by anything associated with the term postmodernism denounces its relativism. Relativism is anathema to scientific inquiry, for it would presumably allow two competing and incommensurable theories to coexist without the ability to differentiate their scientific validity and credibility (e.g., Laudan 1990). If science has made any contributions, these contributions have been commonly associated with the ability to distinguish between belief and truth, between the imagined and the real, between trust in someone's statements and evidence that these statements ought to be trusted. For example, a standard comparison is between magic and science, two areas of research and practice that are supposedly clearly distinguishable using the standard modernist criteria of demarcation between science and pseudo or non-science. Once relativism is adopted, the criteria for such a distinction have been obscured or have changed, thereby suspending the prestige attributed to science and technology as discourses and arenas of practice whose criteria of legitimation are stable and secure. As Charles Newman claims: "Post-Modernism is an ahistorical rebellion without heroes against a blindly innovative information society." (Newman 1985, 10)

Another, related objection has to do with the alleged vacuous nature of postmodernism when one tries to apply it to political practice or technological experimentation. Having robbed science and technology of their esteemed status and materiality—data are not "real" but constructed—and having opened their quarters to anyone and anything that wishes to be called science and technology, postmodernism engenders total chaos and the futility of appeals to stability, security, and even common sense. Science has been traditionally the refuge of brave intellects from religious and superstitious speculation and persecution (claiming its victims along the way, like Giordano Bruno in 1600). Postmodernist views of science demolish this last refuge, leaving nothing, not even common sense, protected from power plays and abuse of authority (e.g., Foucault 1970). One expression of this problem is well summarized by Felix Guattari

(1986) who voices his concern with the politically and materially vacuous structuralist analysis perpetuated by postmodernists/poststructuralists, an analysis that yields no platform or agenda for change. From this perspective, then, being vacuous means being apolitical, which in turn means being socially and morally irresponsible.

From Science to Politics through Critique

Postmodern technoscience is a political mode of discourse when it tries to apply its epistemological concerns to specific situations in which segments of the scientific community are involved, as might be gleaned from the case of the SSC. A postmodern orientation to the study of science may have little to do with challenging, for example, current views on quantum mechanics. Yet, in its insistence on continued forms of critiques, it can encourage continued criticism of and experimentation with the construction and acceptance of the latest "received view." The criticism could come from the meekest voice and from the most remote of locations; yet this could be criticism that displaces one theoretical framework with another. It may seem by now, then, that all we need is some level of tolerance, and that postmodernism contributes to make this possibility come true. However, as will become clear below, the appeal to (liberal) tolerance loses its force as one evaluates the impact of different critiques and the competition for pronouncing the "final word" or "verdict" on the future of philosophy of science, philosophical inquiries in general, or the enterprise called technoscience. In this respect, then, tolerance may be a necessary but definitely not a sufficient condition for the development of critiques of technoscience.

Late twentieth-century critiques of science diverge in their methodological focus and political goals. Among them one can count sociological critiques (e.g., Latour & Woolgar 1986), those of Marxists (Aronowitz 1988) and feminists (e.g., Harding 1986), and even those of the philosophical establishment, from Popper to Kuhn. As we approach the next century, different intellectual scenarios can be envisioned in light of these critiques; they may occur sequentially or simultaneously. First, each mode of critique will continue its efforts to undermine traditional presuppositions concerning science without regard to other critiques. As such, each effort will ignore deliberately those of others. Second, noticing the efforts of others, each mode of critique will attempt to ascend to a position of hegemony from which all others are perceived as either contributing or incidental. Eventually, one mode will dominate and overshadow all others, in the sense of excluding all others from making separate contributions. Third, the rivalry among different modes of critique will cause a disarray of efforts so that their effects will be marginal or overlooked systematically.

When the three scenarios occur simultaneously, it is impossible to get a clear picture of the intellectual scene. Yet, under all three scenarios, one possible outcome looms heavily: the ineffectiveness of any and all modes of critique of science to change anything, so that the status quo is maintained. A plea for some engagement among the different modes of critique is a political plea (in the academic and intellectual contexts) to preserve the possibility of changing not only specific practices of scientists but also the entire political structure of the capitalist environment in which they operate. Dialectically speaking, one possible site for such an engagement may be the conversation of postmodern philosophy of science and postmodern technoscience. With this in mind, I would like to see more clearly how conversations are conducted between Marxists, feminists, postmodernists, Popperians, and any other group of critics that is committed to radicalize the practices—linguistic and institutional—of technoscientists. It should be noted that the conversations I have in mind are heterogeneous practices that defy the uniformity associated traditionally with political or ideological alliances as such.

If the "landscape" portrayed above makes sense, that is, if the three scenarios could possibly take place, and if the plea for critical engagement on strategic grounds is deemed worthy, then it is useful to pay special attention to the form of feminist critique enunciated within the particular work of Donna Haraway. Parts of her work illustrate how one can at once employ both "external" and "internal" critiques, the former associated primarily with philosophers, historians, and sociologists of science and the latter associated with scientists themselves. The dialectical twist undertaken by feminists (along Marxist lines) conflates these two classifications of the notion of critique (internal and external) and claims that it is essential for a critique to be at once immanent to and outside of the practices and theories of science in order to have a perspective that may be blurred enough so as to transcend traditional, oppositional binaries.

A dialectical critique attempts to understand the internal structure of a theory or model or mode of behavior. This understanding—a deconstruction in contemporary parlance—sets the stage for a reconstruction or reconfiguration that may be pushed simultaneously in two directions. On the one hand it provides an opportunity to examine some implicit or tacit presuppositions in an explicit fashion, as for example, the feminist illumination of gender-based forms of discourse and practice; on the other hand it suggests ways in which the theory or model or mode of behavior can be pushed to its "logical extremes" in order to evaluate potential collapses (as in the case of the capitalist mode of production in Marxist terms).

Once the critical process of evaluation is under way, there is also the possibility of setting up alternatives to the theory or model or mode of behavior. Historical examples abound. One example encompasses the debates between the inductivists among the Vienna Circle members and Popper. The inductivist or verificationist agenda was criticized for its inappropriate demarcation of science from non-science. For Popper, falsification is supposed to be the guiding principle for testing empirically the scientific content of conjectures and hypotheses. Did the Popperian principle of falsification (with its attendant notion of verisimilitude) replace the inductivist principle of empirical confirmation? Is his critique replete with a discussion of the metaphysical commitments of scientists—e.g., a belief that the Truth can be revealed once and for all? Or, as some critics continue to claim, has the Popperian critique been usurped by the probabilistic inductivist framework and thus rejected, at this point, as an alternative? Regardless of one's interpretation of the record, the dialectical tension that ensued after the original confrontation has opened up the debate.

Another example is Kuhn's view of paradigm shifts and the relation between normal and revolutionary science: does it *replace* traditional views of the methodology of science (theories devoted to a cumulative and continuous growth of scientific knowledge)? Kuhn's ideas are supported by his interpretation and reconstruction of the historical record, as well as by his focus on certain sociological aspects that organize and influence the growth of scientific knowledge within the scientific community. Kuhn's critique, then, attempts to recast science in a different light, one presumably deemed more accurate or appropriate. But Kuhn's critical success has become, almost immediately with its introduction, a target of relentless critiques: some charging his views with the irresponsibility that accompanies irrationalism, while others claiming that his view is so vague and ambiguous that it is no coherent critique at all. Whatever the particular agenda of these opposing critiques, they have helped develop and expand the debate on the internal functioning of the technoscientific community.

Marxist critiques of science, from Karl Marx (1844) to Stanley Aronowitz (1988), share a common thrust in that they expose science for what it really is: a capitalist and bourgeois activity whose goal is the exploitation of the working (and probably, by now, also the middle) class. This critique of science tries to denounce the mystification surrounding scientific jargon and texts in order to reveal the ideals of truth underlying them. Contemporary Marxist critiques are at times lumped together with feminist critiques (from Keller to Harding and Haraway), though they use different discursive techniques and have different objectives in mind. Overlooking for the moment the great divergence of techniques and ideals, it is clear from these critiques that they wish to replace scientific practices prevalent today with some other, better forms of discourse and practice. As such, they all contribute to the ongoing debates over the context within which technoscientific practices are performed.

Are postmodern critiques different? Dialectically speaking, the answer is at once yes and no. Just as most Marxist critiques hold onto banners, such as exploitation and alienation, as useful terms with which to anchor their discourses, postmodern critiques wave, among many other banners, that of displacement. This term is supposed to replace (with a twist of irony) the term *replacement*, so that the postmodern discourse should be accorded residency alongside or next door to other discourses, without thereby pretending to surpass or overcome previous discourses in the Hegelian sense of *Aufhebung*.

As someone who is not "part of" the feminist discourse, but nevertheless regards himself a reasonable listener and translator, I find interesting parallels and points of intersection between feminist and postmodernist critiques of science. But that is not enough. Having said earlier in this chapter that feminist critiques are essential in order to make sense of postmodern critiques of science, one may recall Suleiman's warning about the apparent "opportunism" that may undermine an effort such as mine in this context. That is, one may dismiss my attempt to listen to feminist critiques as a fetishized attempt to hook onto the latest intellectual fad or fashion. Yet Suleiman reassures me by saying, "There is, I believe, an element of mutual opportunism in the alliance of feminists and postmodernists, but it is not necessarily a bad thing." (Hoesterey 1991, 116) If I understand her correctly, Suleiman acknowledges the possibility of forming strategic alliances that would have a stronger political force to change the "established" or canonical view of technoscience.

It is not only interesting but necessary for postmodern philosophers of science to consult the work of feminists. For it seems that the canonical spokespeople for postmodernism (e.g., Derrida or Deleuze and even Lyotard) shy away from considering science and technology in the manner that Haraway does. When they in fact mention science and technology, their critiques fall short of providing the sort of analysis that they provide regarding architecture or literature (e.g., Lyotard maintains a certain level of deference to scientific "descriptions," see Sassower & Ogaz 1991). Feminist critiques, therefore, lead the way for an improved version of a postmodern critique of science. Without them, postmodern philosophers of science will resort to their old examples, just as I have done above, concerning the Vienna Circle and Popper or Kuhn and Feyerabend.

Feminist and Postmodern Engagement

In this section I only hint at the potential engagement between some feminists, like Haraway, and some philosophers of science in the Popperian mold; a fuller explication is presented in Chapter 5. What is important to mention here is the political tone of Haraway's critique. She confronts discursive practices in the following manner:

> The evidence is building of a need for a theory of "difference" whose geometries, paradigms, and logics break out of binaries, dialectics, and nature/culture models of any kind. Otherwise, threes will always reduce to twos, which quickly become lonely ones in the vanguard. And no one learns to count to four. These things matter politically. (Haraway 1991, 129)

For Haraway, the allure of postmodern pluralism is not reduced to an engagement in the creation of different vocabularies for the sake of it. Rather, different vocabularies will admit a plurality of alternatives and answers to any question, such that no single answer will ever be sufficient for an audience. With more than one alternative at hand, there is a built-in dialectics at work, and no claim for hegemony can be fully defended.

In this context, then, there is an appeal to multiple options that in their very formation bespeak heterogeneity (Ibid., 193). Heterogenous multiplicities reflect the concern with and the ability to translate the postmodern commitment to pluralize epistemology into "an argument for *pleasure* in the confusion of boundaries and for *responsibility* in their construction." (Ibid., 150) So Haraway's concern with confusion, unlike the game metaphor used by postmodernists, is a concern with responsibility (more on this in Chapters 5 and 7). Put differently, whereas some postmodernists have been accused of being irrational or irresponsible, advocating what seems to be intellectual chaos, some feminists, such as Haraway, insist on the responsible construction of blurred boundaries in the aftermath of deconstruction.

In Haraway's construction, then, "Some differences are playful; some are poles of world historical systems of domination. 'Epistemology' is about knowing the difference." (Ibid., 161) Epistemology, according to her, takes on a different form of orientation, one that no longer merely "plays" with the latest data or reconstructs a new paradigm that would revolutionize scientific thought, but instead contextualizes critical judgments and affirms particular political commitments concerning any contestant for the status of a paradigm.

Offering a different vocabulary may not resolve, but may rather problematize the choice between replacement and displacement of one vocabulary with another. Haraway does not wait for another to recognize this problem, but problematizes it herself: "The feminist dream of a common language, like all dreams for a perfectly true language, of perfectly faithful naming of experience, is a totalizing and imperialist one." (Ibid., 173) In order to avoid becoming one more stage in some universal (Hegelian) system, she recommends perceiving the feminist critique of science as engaged in "cyborg politics," one that is about "noise." (Ibid., 177) Her insistence on noise as opposed to a unified, clear, and totalizing language would ensure that a feminist critique is not about an anti-science that rebuilds dualisms (Ibid., 181). She reminds her readers that "Science has been utopian and visionary from the start; that is one reason why 'we' need it." (Ibid., 192)

When Haraway explains her epistemological—read political—viewpoint, she enters the debate about "Situated Knowledge," a debate that promises to supersede binaries (Ibid., 187). According to her:

> Situated knowledges require that the object of knowledge be pictured as an actor and agent, not a screen or a ground or a resource, never finally as slave to the master that closes off the dialectic in his unique agency and authorship of "objective" knowledge. (Ibid., 198)

What differentiates her posture on this question from Baconian and Kantian attempts to appreciate objective knowledge in intersubjective terms (as was the accepted norm of the Royal Society of London) or from the attempts of pragmatists to contextualize knowledge in practical, goal-oriented frameworks is her sense of responsibility. She says:

> The moral is simple: only partial perspective promises objective vision. This is an objective vision that initiates, rather than closes off, the problem of responsibility for the generativity of all visual practices . . . Feminist objectivity is about limited location and situated knowledge, not about transcendence and splitting of subject and object. In this way we might become answerable for what we learn to see. (Ibid., 190)

The answerability of feminist critiques rests on a redefined notion of rationality: "The science question in feminism is about objectivity as positioned rationality." (Ibid., 196) "Positioned rationality" could be paralleled to the critical and relative rationality about which Agassi and Jarvie speak (1987) or with the one associated with postmodern forms of contextualization (Ormiston & Sassower 1989). Chapter 5 is devoted to the examination of this potential engagement. But, as was already mentioned in the case of postmodernism in general, can Haraway's view escape the standard criticism that it falls into a form of irrationalism or relativism?

Her response is: "The alternative to relativism is partial, locatable, critical knowledges sustaining the possibility of webs of connections called solidarity in politics and shared conversations in epistemology." (Haraway 1991, 191)

Liberal Intolerance and Radical Engagement

"Webs of connections" can be understood as "labyrinths," wherein it is difficult to distinguish one narrative from another because each is linked in more than one way to every other (Ormiston & Sassower 1989). Yet, various differences may remain between these two metaphors or images, wherein the former reflects an intimate interweaving of sets of ideas and discourses, while the latter presents the possibility and not necessity of interconnections. Systematically deconstructing any claim for singularity and uniqueness may pose the problem of relativism, as Haraway recognizes. Her response is the insistence on "shared conversations in epistemology" and "solidarity in politics." Does Haraway promote the obliteration of differences? Unlike Rorty's liberal position, hers does not push for this sort of alternative.

The "S" word (solidarity) has been used extensively by Rorty. Earlier I claimed that Rorty would appreciate my reading of Bacon in a postmodernist vein. But Rorty the postmodernist turns out to be a liberal of a particular kind, perhaps bourgeois, perhaps privileged. For example, he speaks of "human solidarity" (Rorty 1989, xv)—understood in terms of a "liberal utopia" (Ibid., 190)—in his interpretation of the postmodern orientation that includes "contingency" and "irony." But Rorty's solidarity is not Haraway's, and this difference is important to note if we are to differentiate between the critical engagement of postmodernism and technoscience and that of liberal modernism and science.

Though Rorty is quick to invoke the notion of utopia in his work, it is clear it is not the utopian vision of Marx or that of Haraway. For him the utopian dream of the "liberal West" is of "Tolerance rather than that of Emancipation." (Rorty 1991, 213) This may sound at first like my own minimalist position that would get disputants to talk to each other, but closer reading shows that a strong liberal foundation supports his view of tolerance. Revolutionary politics, for Rorty, is "no more than intellectual exhibitionism," an activity that is "not interested enough in building causeways" to connect intellectual "islets" with the "mainland." (Ibid., 221)

Rorty positions himself against the backdrop of French philosophers and does not directly address in this context feminist critiques of science. But if the political alliance I try to establish between postmodern and feminist critiques of science came to fruition even temporarily, Rorty's posture would undermine it at once, antagonizing feminists along the way. Rorty's views are relevant in this context, for he can easily divert any effort to radicalize technoscientific discourses (at least in the sense of insisting on heterogeneity when homogeneity has been the only acceptable standard) to a liberal fold that fails to see the stakes feminists, for instance, have in critically engaging science and its practitioners. This is not to say that there are no liberal feminists or that there can be no engagement between some liberals and some feminists.

Yet, I would like to point out what I find most problematic in Rorty's assessment of the status of postmodern and feminist critiques: his cavalier use of the term "mainland." The solidarity he seeks is with the mainland—that is, the liberal intellectual establishment—one that is comfortable, as he says, with "splitting the difference between Habermas and Lyotard, of having things both ways." (Hoesterey 1991, 94) But some differences are incommensurable, and sometimes differences cannot be simply "split": the truth may not lie somewhere in the middle!

Haraway's notion of "conversation" (see also Haraway 1991, 201), parallels my sense of critical translation or radical engagement, where disagreement is not glossed over nor necessarily mitigated. Postmodern technoscience can be the forum for such an engagement, where critical evaluations and radical utopias would be solicited continuously. My view differs from Rorty's because I no longer expect that conflicts need to have causeways to Rorty's mainland; even when causeways are used, there is no presumption that there ever is a mainland whatsoever, only various islands from and to which we may travel. In its maintenance of some utopian ideals, my view agrees with Peter Koslowski that "Postmodernity postpones the final decline [*Untergang*] that is supposed to occur after the collapse of the utopian expectations contained in modernism's philosophy of history." (Hoesterey 1991, 146)

Rorty's tolerance turns out to be a liberal *in*tolerance toward the continued critical and radical engagement his views may suffer from postmodern and feminist quarters, that is, from those who still dream of challenging the stability and hegemonic power of the technoscientific discourse and community. Having dismissed emancipation for the sake of tolerance, Rorty betrays his own dominance (for which no emancipation is needed) and paranoia (pleading for tolerance so that his position of power will not be viciously attacked). Perhaps Rorty's dichotomy of tolerance and emancipation must be overcome so that an emancipatory utopia is tolerated by liberals as well as upheld by radical critics.

When some feminist critics of science devise strategies to undermine the hegemonic position and status accorded to science, they use epistemological arguments laced with political and social concerns. Alleged postmodernists such as Rorty can inadvertently undermine and derail the efforts of other postmodernists and feminist critics of science, because they fail to pay attention to the psychological settings that enhance or retard possible discursive transformations. The psychological dimension I wish to add to the political and epistemological concerns already mentioned by others tends to the personal commitments and convictions of individuals, their stakes in advancing critiques, and the personal risk and potential empowerment, if not emancipation, they may suffer or enjoy respectively as a consequence of these critiques.

So in some sense, I wish to suggest a strategy for action that is simultaneously another critique of philosophy of science and technoscience and a translation from one critique to another. My strategy recognizes the need to refer to more than one critical discourse or literature at any given time. This recognition reflects discursive and practical heterogeneities. In doing this, I stand on the shoulders of many giants (see Merton 1985), perhaps too many. This posture has the accompanying danger faced by any acrobat in the carnival of the end of the twentieth century: bouncing between and then falling off these shoulders. If I do fall off, I hope it is a soft landing and not a landing wherein the frenzy of carnival participants tramples me to death.

Chapter 3

Translation as a Political Agenda

Introduction

Claiming that everything is political is at once claiming too much and too little. If everything is political, then the unique political dimension of particular acts and modes of interaction and exchange fails to illustrate the special warrant for its designation as political. Yet it is worth assigning a value to an activity or a thought, a theoretical treatise or an exchange, for such a value may highlight what is already implicit. That is, I understand my own undertaking in this book—my translation—to be political through and through, and this chapter will attempt to explain why this is indeed the case. But to claim a political stand for one's posture, as I did in Chapter 2 in relation to Rorty, is to acknowledge that one's own declaration is self-legitimating. In this respect, then, and with a clear view of the different postmodern gestures made over the past two decades concerning the status and legitimacy of one's own words, I suggest that my political posture, my translation, be appreciated well beyond the confines of the Marxist or the Critical tradition of the Frankfurt School.

Some could argue that the sheer declaration that one's work is Marxist provides sufficient grounds for assuming all that needs to be assumed, without arguing for anything in particular. Similarly, some would argue that when people consider themselves liberal, this should give sufficient indication to all who care to know not only the genealogy of their views but also the policy implications of these views and commitments. To some extent this makes sense: nothing is more efficient than codes and symbols, shorthand devices by which to encompass a range of issues and historical precedents so as not to repeat oneself and reinvent the wheel. It should be enough to say that you are a liberal or conservative in order to appreciate the full range of convictions associated with these positions. But as any novice semiotician will protest, there are layers of meanings that do not reveal themselves on the face of a declaration or a word. Instead, a hermeneutical moment or a deconstruction is needed, as the Talmudic students of the past two millennia have known all along. Signs ranging from divine messages to computer printouts invite a multiplicity of interpretations. This situation in and of itself should offer advantages to liberals and radicals alike, for it enhances popular participation in the activity of interpretation. Everyone should feel invited to contribute something personally valuable to a seemingly objectified and detached sign.

However, when one wishes to know the "correct" or "right" interpretation, a problem inevitably arises: whose interpretation counts, and whose is the correct one? What criteria are to be used by all so that only one correct interpretation is chosen? What happens if two competing interpretations are put forward? Which one should be chosen, and now by what (different) criteria? Furthermore, if one interpretation is chosen "above" all others, what status does it have, and what status is rendered to the others? Is one interpretation, in fact, above or superior to all others? And, most importantly for my context, even if we all agree on a single interpretation that supersedes all others, is its superiority temporary or permanent?

From a postmodern perspective, one that is articulated in each of the chapters in this book (depending on the particular context of each discussion), I take it for granted that one is always bound to make a pragmatic or existential choice, and that that choice is always temporal. This does not mean that the choice is by definition meaningless or haphazard. On the contrary, when the choice is made within a particular context it holds the authority rendered to any choice: it is valued critically and respected. This does not mean that no challenge can be mounted to replace or displace the choice. A *choice interpretation* remains authoritative in many instances because it is difficult to find the means by which to challenge it or because the criteria according to which the choice was made remain either implicit or too entrenched to be re-examined. To ensure the temporality of choice interpretations, the conditions for challenge must be underscored. What can be done to the conditions of challenge so that they encourage rather than retard potential challenges?

First, it should be clear that although the challenge proposed here may seem limited to theoretical activities or contexts, I hope, as later chapters will illustrate, that this is not the case. The challenge I propose is simultaneously theoretical and practical, thus collapsing the need to uphold this age-old dichotomy into praxis.

Second, as for changing the conditions for challenge, the change can come about either by a careful examination of these conditions from within or outside the context that brought them about. If the approach is "internal" to the context that gave rise to these particular conditions, there may be a "reform" of these conditions. On the other hand, if the approach is "external," a "revolutionary" transformation of these conditions may lead to a revolutionary challenge and a different choice interpretation. I have used loaded terms, such as internal and external, reform and revolutionary, on purpose, so as to demonstrate the layers of meanings and presuppositions that are brought into play whenever one begins to discuss the potential for challenge. These terms, then, are at once descriptive and prescriptive, sociological and political, philosophical and linguistically potent.

Third, there may be a difference in approach toward the enhancement of a challenge to the status quo, to the choice of a "preferred" interpretation of the data, the political parties, the scientific model, or the healthcare policy. Assume, for the moment, that the approach is external to the context under which a set of criteria for choosing among alternative interpretations were formulated. How does an external approach relate to the internal workings, here not only sociological but linguistic, of the particular community or discourse? How are connections made? What form of communication must be assumed or taken for granted in order to relate criteria, context, interpretation and the wholesale critique of these three sets of variables?

These questions lead me to think about the role of philosophers and the means by which that role can be performed, as I intimated in the Preface. As far as I can tell from my own work, philosophers ought to be cultural critics, and they may fulfill this role most productively by being translators. Let me explain this point slowly.

Politics of Translation

The (in)famous Thesis XI on Feuerbach says: "The philosophers have only *interpreted* the world, in various ways; the point, however, is to *change* it." Marx admonished his fellow philosophers to do something significant with the tools of their trade, their intellectual weaponry, to use their skills and power for the sake of creating a better world, a world devoid of suffering and pain, exploitation and hunger. Marx's admonition remains alive today, no matter what political or ideological label one chooses. For if philosophers refuse to change the world, they contribute to their own marginalization, despite the prestige and authority granted to them at times as members of the academy.

Marx says change is imperative. I say translation contributes to bringing about change. Therefore, I believe it is a reasonable strategy for intellectuals, philosophers included, to participate in different modes of translation. They should all undertake the travels of foreigners even within the confines of their own academy. They should always engage the anxiety of outsiders as they continue their activities inside.

Now, what form will their change take? Biblical prophets and Greek oracles cited divine inspiration as the basis for their judgments on the affairs of the state. They shielded their critiques and themselves from otherwise certain annihilation by claiming divine protection, by appealing to a power that could not be overcome by mere mortals. Stalin's sarcastic remark during World War II about the "power" of the Pope ("how many divisions does he have under his command?") notwithstanding, there has always been a refuge for critics, whose power has not been measured by earthly standards. So, to begin with, I guess it seems reasonable (if not rational) to introduce a genealogical link between oracles and prophets, sages of the ancient world, and contemporary philosophers. This link is, of course, ideal in the Weberian sense, and is used as a heuristic device with which to articulate the role of philosophers and then examine the means by which they may fulfill this role.

John Locke's minimalist definition of the philosopher's role as the "under-labourer" of science focuses on the relationship between scientists and other intellectuals, rather than on the relationship between monarchs and philosophers. He therefore seems to view philosophers not as critics but as apologists, that is, as those who clarify rather than obscure linguistic ambiguities. In his words:

> The commonwealth of learning is not at this time without master-builders, whose mighty designs, in advancing the sciences, will have lasting monuments to the admiration of posterity; but everyone must not hope to be a Boyle or a Sydenham; and in an age that produces such masters as the great Huygenius and the incomparable Mr. Newton, with some other of that strain, it is ambition enough to be employed as an under-labourer in clearing ground a little, and removing some of the rubbish that lies in the way to knowledge; which certainly had been very much more advanced in the world, if the endeavours of ingenious and industrious men had not been much cumbered with the learned but frivolous use of uncouth, affected, or unintelligible terms, introduced into the sciences, and there made an art of, to that degree that philosophy, which is nothing but the true knowledge of things, was thought unfit or incapable to be brought into well-bred company and polite conversation. Vague and insignificant forms of speech, and abuse of language, have so long passed for mysteries of science; and hard and misapplied words, with little or no meaning, have, by prescription, such a right to be mistaken for deep learning and height of speculation, that it will not be easy to persuade either those who speak or those who hear them that they are but the covers of ignorance, and hinderance to true knowledge. (Locke 1964, 58–59)

According to Locke, the ideal speech situation is one wherein true knowledge is associated with philosophy, so that the great insights of science are not obscured but rather made transparent. The use of words and terms should enhance our understanding and bring us closer and closer to true knowledge. The role of the philosopher, then, is to clear the ground surrounding the great monuments built by scientists. Locke's image is that of the citadel of learning, the ivy tower, the site for which "master builders" are responsible. It is not only visible to their contemporaries, but its monumental design will be saved for and admired by posterity. The "rubbish" that bothers Locke is linguistic; it is not the rubbish of the streets of London, which infects people and kills them, or through which plagues are transmitted to entire cities and continents. The world Locke cares to change and improve is the intellectual-*qua*-scientific one, and not the world of oracles and prophets. Their ancient world was infested with monarchs and villains, monsters that threatened the well-being of innocent and law-abiding citizens. They cared about justice and fairness, while Locke cares about clean, true knowledge. What does Marx care about?

If we take Thesis XI seriously, it would seem that Marx cares more about justice than about clean knowledge. But this reading is highly problematic, for Marx and his disciples argue that in order to appreciate justice and bring it about under the material conditions of the state, one must have "true" knowledge of these conditions. In this sense, then, one must not pursue true knowledge *or* justice, but true knowledge *and* justice. It is with this interpretation of Marx's view of the role of philosophers that I explain my contention that the role of philosophers is to be cultural critics. I should add that the term *culture* encompasses the entire web of social, economic, political, and ideological conditions under which personal decisions and public policies are made. The field of Cultural Studies exemplifies some of the complexities and excitements associated with this concept (e.g., Best & Kellner 1991).

Cultural critique is not limited to Locke's concern with language and its "abuse." Such a constraint would restrict the philosophical tradition to its dominant recent guise as "analytic philosophy," a tradition some of whose practitioners admit is confined to the role of the gatekeeper of logical inconsistency and is therefore at a dead end. Instead, I find cultural critique most usefully articulated with the aid of a multiplicity of sources and discourses. That is, instead of narrowly pursuing the clarification of cultural terminology and grammar, I try to learn a *variety* of terminologies and grammars so as to translate between discourses. But translation requires both familiarity with a variety of discourses and the possibility of translation.

It is important to appreciate that different discourses not only use different vocabularies, but are articulated and proliferated for different purposes. For example, some contemporary anthologies have strategically posed a topic for discussion, such as "chaos and order," and have asked scholars from different academic disciplines to explore these concepts from their own perspectives, from literature to science (Hayles 1991). The process of translation is not as simple as it may sound at first, nor can it be achieved simply by reading the texts of others. Perhaps what can motivate and enhance the process of translation is a general sense of injustice or a shared attempt to resolve a particular problem; in this way the rationale for bringing together a multiplicity of discourses can become clear and reasonable.

As an example, it is possible to consider a particular set of discourses whose texts and the many interpretations and elaborations they have spun over the years are considered quite incommensurable: Marx, Popper, Lyotard, and Haraway (in chronological order of the appearance of their texts). Is a translation between them possible? Is it fruitful? Popperians do not incorporate the work of postmodernists and feminists, just as postmodernists are unlikely to incorporate the insights of Popperians and feminists, and the same goes for the work of feminists and Marxists who ignore Popperians and postmodernists *and* each other. Every once in a while an appeal is made to incorporate a variety of discourses and assimilate the "best" each has to offer (e.g., Nicholson 1990), but such an appeal falls on deaf ears more often than not, perhaps for the simple reason that it has become impossible to keep up with the proliferation of the different literatures marketed throughout the academy.

Marxists can hardly keep up with their own factions, and the same is true of postmodernists and feminists; perhaps the Popperians fare a bit better because they constitute a small group, but since their motto is rational criticism, it is unimaginable that they will ever solidify a consensus for a uniform viewpoint. It is obvious that despite the trivial glaring differences between these traditions or schools of thought, there are good reasons for crossing their discursive boundaries and learning from each other (Sassower 1993a, Ch. 6). How is that to be done? My answer is through translation. Philosophers have the burden of making that translation possible by reading across discourses, learning enough of the vocabulary and grammars to repeat, in different words, the meaning they cull from the depths of these discourses. This is no mean trick.

Think for a moment what it would take to paint a picture or set up a grid of the multiple answers given by some Marxists, feminists, postmodernists, and Popperians on the following questions: a) what view of truth do you hold? b) is objectivity possible? and c) what is science? The grid would have twelve squares for the answers. Add to that my own self-reflexive move, and three additional answers will be given. At this point the reader is faced with comparing fifteen answers for differences and similarities. This grid, incidentally, could be set up as the table of contents for an entire book.

But what about the terms used by each school of thought? Does Lyotard mean by "objectivity" exactly what Popper does? And what about Haraway or anyone else who is chosen as a representative? Moreover, can the individuals chosen as spokespersons be granted the status of representatives? It is not only a question of whether or not they represent accurately or fairly or comprehensively the views of their colleagues, but rather whether or not they can represent them at all, as we have already seen in Chapter 2 in the case of Rorty, the alleged postmodernist. Hegel alerted us to false consciousness in his phenomenological study; Marx pushed this point to a greater social and scientific extent; and finally, Freud tried to convince us all that it is impossible to represent even the most basic and personal issues regarding one's own personality (unless aided by a therapist or psychoanalyst). So, we are in a bind!

In typical philosophical fashion, I have managed to establish enough obstacles on the way that the reader may doubt that the reading of any subsequent chapter will lead anywhere. So as I plead for the opportunity to continue my exploration, I shall have to make some concessions, employ conventional definitions, and assume a whole range of issues in order to expedite the journey proposed here. To begin with, I will assume that some sort of representation is warranted in the cases examined in subsequent chapters, so that Haraway's work, for example, is in fact used fruitfully to exemplify a whole range of feminist concerns that may be articulated differently by her colleagues.

As for the terminology used by different authors: whenever necessary (for my translation purposes), I define and explain the terms and examine their different proscribed meanings. This, then, would be part of the work of translation undertaken here. But no matter how philosophically careful I attempt to be, it should be recognized and acknowledged that a certain level of indeterminacy inevitably pervades the entire book, much to the chagrin of those who wish a clean-cut solution to or presentation of the interconnectedness of multiple discourses. Furthermore, I wish to reassure the reader that the authority I assume in my own form of translation is temporal and partial; it is the authority of a foreigner who depends on the generosity of hosts.

Besides, any one of the terms used here, such as truth, objectivity, or reality, may be understood by some as explicable only within the domain of classical epistemology. But what if the articulation is extended beyond epistemological confines? For some, epistemological choices pertaining to objectivity are based on "socially shared plausibility judgments rather than proof" (Campbell 1993, 99). For others, truth-claims move from the social to the ethical: "Truth is usually regarded as a category of epistemology . . . it has all the qualities of an ethical ideal, and indeed that it is easier to defend as a category of ethics" (Collins 1993, 303). This analysis of the scientific community and the ideals it both upholds and contests are understood in Durkheim's sense and therefore take on this shade of analysis. But can this move be taken seriously by traditional philosophers of science?

Indeterminacy of Translation

Among American philosophers who have considered the implications of the debates inspired by Ludwig Wittgenstein and carried out during the two World Wars by the Vienna Circle is Willard Van Orman Quine. His echo of Otto Neurath's concern for the philosopher's role suggests a position between Locke and Marx: "The philosopher's task was well compared by Neurath to that of a mariner who must rebuild his ship on the open sea." The ship he talks about is not unlike the monument mentioned by Locke, for it is a conceptual scheme that stands for or represents reality. Quine continues to carve a middle ground by refusing to fall into either of the ontological extremes of constructivism (reality is constructed by humans) and naturalism (reality exists independently of humans) by proposing his version of pragmatism:

> It is meaningless . . . to inquire into the absolute correctness of a conceptual scheme as a mirror of reality. Our standard for appraising basic changes of conceptual scheme must be, not a realistic standard of correspondence to reality, but a pragmatic standard. Concepts are language, and the purpose of language is efficacy in communication and in prediction. Such is the ultimate duty of language, science, and philosophy, and it is in relation to that duty that a conceptual scheme has finally to be appraised. (Quine 1963, 79)

In this light, then, it is clear from Quine's perspective that what is in fact under scrutiny is not reality as such (for it is impossible to have "direct" access to reality), but rather the languages with which we try to communicate to each other about it. In his words:

> I am not suggesting a dependence of being upon language. What is under consideration is not the ontological state of affairs, but the ontological commitments of a discourse. What there is does not in general depend on one's use of language, but what one says there is does. (Ibid., 103)

It may seem at first glance that ontological questions, as mentioned here by Quine, digress from my concern with translation among multiple discourses on technoscience and scientific knowledge. But they do not; for the use of language presupposes some ontological commitment (or a sense of reality), and, therefore, that commitment needs be made explicit to allow any sort of translation. Quine makes it clear that the indeterminacy of translation is a condition of language and not an aberration, that it is something we have to learn to live with and treat appropriately rather than expect to cure. Quine follows here Benjamin Lee Whorf's view of the organizing function of language, that it helps classify and categorize (as Plato illustrates in the dialogues of Socrates).

Donald Davidson reiterates this position and emphasizes that in fact, language has the same general function as science. As such, it seems to refer to the "same" experience or sense data, but because it employs different conceptual schemes (that eventually bring about different linguistic expressions and formulations), there is a "failure of intertranslatability ('calibration')" (Davidson 1984, 190). Davidson develops Quine's notion concerning the problematic relation between language and reality, between an ontological given and the relative conceptual schemes or languages that organize, describe, or fit it. So Davidson's sense of the failure of intertranslatability and Quine's sense of the indeterminacy of language are extensions of each other: they both point to the relativity inherent in linguistic expressions, without thereby dismissing the importance of translation or reverting back into the postmodern discourse of deconstruction.

While considering "radical translation"—that is, a "translation of the language of a hitherto untouched people" (Quine 1960, 28)—as opposed to a translation between two well-known languages that can be undertaken by a bilingual (Ibid., 47), not to mention a translation between two discourses using the same natural language, Quine concludes that what can successfully accomplish such a translation is the "native's behavior," "non-verbal stimulation," and general "social assessment." (Ibid., 30–32) This would mean, for instance, that although a contemporary discourse may define itself along constructivist lines, its practitioners may still refrain from walking through brick walls or on water.

The notion of translation has a variety of relevant contexts within which it can be defined. For example, one may attempt to translate a word or a sentence into another word or another sentence within the same language, or attempt to translate a word or a sentence from one language to another. In either case, there are additional conditions to be satisfied for the translation to be a good as opposed to a bad one: the meaning and the reference should be preserved. Even this oversimplified characterization alludes to a third party, so to speak, according to which one may test—confirm or falsify—the value of a translation: the natural world or reality. We are back to Quine's ontological discussion or to the basis on which claims for scientific knowledge are commonly made. (More on this in Chapter 6.)

If we have two (natural) languages that describe similar phenomena, it would make sense that in principle one could successfully translate terms and sentences from one language to another, perhaps with the aid of specific reference points. But, as Quine says, "reference *is* nonsense except relative to a coordinate system." (Quine 1969, 48) The problem, of course, is that of reduction, that is, of reducing all sentences to empirical sentences and then formulating them as so-called protocol sentences so that they correspond directly to sense data. This exercise was tried by Neurath and Rudolf Carnap and their colleagues in the Vienna Circle and then abandoned partially (Ayer 1959), for it leaves too many expressions and verbal communications outside the range of acceptable meanings. Moreover, there are problems associated with the choice of a coordinate system that somehow relates to one's "mother tongue" so that its words are taken "at face value." (Quine 1969, 49) There are also problems concerning the legitimation one renders to one theory in the name of science, for there ought to be a level of possible reduction from one set of sentences to another that does not result in a radical reduction (Bunge 1991).

The abandonment of radical reduction of words and sentences into empirical observations allowed for these related developments in the analysis of language. First, it was acknowledged that there could be meanings not reducible to empirical data, such as metaphysical statements and speculations (e.g., Popper 1959, Ch. 1). This allowed for a greater appreciation of the rich semantic basis of words and sentences that transcends the confines of the rigid rules of syntax.

Second, it paved the way for a recognition, as expressed by Quine, that "ontology is indeed doubly relative. Specifying the universe of a theory makes sense only relative to some background theory, and only relative to some choice of a manual of translation of the one theory into the other." (Quine 1969, 54–55) This undermines any belief in the absoluteness of the ontological basis of languages, and thereby complicates the results of translations between two languages, both of which stand on shaky ontological ground.

Third, it was likewise acknowledged that languages are not reducible to the syntax of their constitution, so logical connections and quantifiers, for example, are important devices by which to ensure non-contradiction, but they are not the means to enforce a single meaning of a sentence. According to Quine: "The division between the words that are to be viewed as referring to objects of some sort, and the words that are not, is not to be drawn on grammatical lines." (Quine 1960, 18)

Fourth, natural languages are social phenomena used for communication whose life and changes cannot be regulated by logic alone. In Quine's words: "Beneath the uniformity that unites us in communication there is a chaotic personal diversity of connections, and, for each of us, the connections continue to evolve." (Ibid., 13) That is, conventional meanings are attached to certain expressions and words for the sake of daily communication, so it is necessary to provide a context for defining words. The context at times is the particular sentence in which words are placed, and at times it is a set of sentences, or entire portions of the language.

From these four points, it seems that translation of a word or sentence from a foreign language may require at times an understanding of the entire language, thereby taking into consideration a web of sentences whose inter-relatedness may elucidate a meaning not readily apparent in one word or even in the sentence in which it is used. As anthropologists, such as Peter Winch, continue to argue, the difference between the so-called scientific approach and the approach of magic depends on the difference between cultures. In this Wittgensteinian respect, concepts and what Quine calls conceptual schemes are examined not in isolation, one by one, but in their full use within a language game (Winch 1964, 307, 309). This situation is not limited to theoretical sentences, because empirically loaded sentences also stand on ontological grounds that are relative to some background theory or language. (Quine 1960, 77–79)

Would it not be possible to perform translations from one language to another by finding logical connections between the logical rules that regulate each different language? That is, would it not be possible to circumvent some of the ontological difficulties posed above by formalizing all languages completely so as to make translation a mechanical undertaking? Here we are again faced with the problem of reduction as well as the problem of the social determination of languages, or what has been dubbed the underdetermination thesis of Quine.

According to Lars Bergstrom, Quine's concern with empirical underdetermination is that two theories "would explain the observable features equally well but in different ways, and neither theory would be reducible to the other." (Bergstrom 1993, 333) Bergstrom maintains that it then becomes a problem to know, not only believe, that one's theory is indeed true if one accepts the underdetermination thesis, for the justification of one's belief can hardly be expected to rest on the shaky grounds of underdetermination (Ibid., 344). This situation brings about a kind of skepticism that may not be acceptable to those concerned with scientific knowledge—linguists and philosophers alike, not to mention practicing scientists.

For example, James Bohman summarizes the views concerning interpretation and indeterminacy as weak and strong holism, ascribing to the former the unwarranted rejection of all knowledge claims and thereby a somewhat radical skepticism (based on too radical a relativism) and to the latter a more acceptable form of rational interpretation that allows cross-cultural comparisons (based on a transcendental argument made by Davidson and Habermas). As social scientists consider the views of Wittgenstein and Quine, as they are confronted with ontological relativism and the lack of appeal to an objective basis from which comparisons and judgements are to be made, they need to find a way to warrant any of their models and theories. (Bohman 1991, Ch. 3)

It is not universally accepted that Quine's view is skeptical, though it is quite generally accepted that his view is conventionalist, at least in the sense already developed by Pierre Duhem around the turn of the century. I mention Duhem because the conventionalist thesis of scientific theories has been named the "Quine-Duhem conventionalist thesis," and it holds that "any statement can be held to be true no matter what is observed, provided that adjustments are made elsewhere in the system." (Presley 1967, 7:54) Duhem is one of the first to intersperse philosophical concerns with the methodology of science and the metaphysical foundations on which it is based, and to add a generous dose of the history of science. His realization of the difficulty of ascertaining the ontological basis of science (for humans are only capable of representing images of reality) leads him to the following conclusion: "the hypotheses of physics are mere mathematical contrivances devised for the purpose of saving the phenomena." (Duhem 1969, 117)

So the conventionalism of Duhem and Quine is one that appreciates the impossibility of the direct mapping of the world onto some theoretical model, and that therefore removes the need to examine theories as wholes. As such, then, it would make sense that individual sentences could be viewed as either true or false, for the conditions under which they relate to the rest of the sentences in the theory ensure their truth-value. Hence, to ascribe a skeptical nature to Quine's underdetermination, as Bergstrom does, is not so far-fetched.

Perhaps it is at this juncture that a postmodern dose of skepticism is in place, that is, Jean-François Lyotard's sense of language games. But before I move to this stage, I should emphasize that I view the different discourses mentioned above—the Marxist, feminist, postmodern, and Popperian—in a way that requires a translation of sorts. I have therefore discussed Quine's concerns with translation, his conclusion that indeterminacy is inevitable in all cases where one searches for meaning and truth, and the skeptical sense that emerges from this relativist conclusion.

Language Games

Recapitulating some of the issues raised in the previous section, let me suggest the following schema, as proposed by Nathaniel Laor. Though concerned more specifically with schizophrenia and the techniques by which to overcome the obstacles of communication between patients and therapists (not to mention the split or multiple discourse of the patient), Laor poses the background situation in the following way:

> The general problem of translation has permeated the discussion in both linguistics and philosophy, especially since translation is a prerequisite for rational argument and understanding across linguistic boundaries and intellectual systems. (Laor 1990, 145)

According to Laor, there are two extreme positions concerning language that define a continuum of other positions. At one end is Quine's (conventionalist) view of the social determinants that constitute a speech community within which there is an agreement on a set of verbal rules. At the other end is Noam Chomsky's (naturalist) view of the deep structures that underlie all languages, such that once a reduction to that level is achieved, there is no problem finding agreement among all those who communicate with each other. Between these two views, Laor positions his own view, one that follows Johann G. von Herder.

Laor begins by admitting, in a manner no different from that of other linguists, that it is unreasonable to assume that language is by definition transparent. Instead, he claims that "language is always opaque and metaphorical" and continues to say that it is therefore "fallible and inadequate." (Ibid., 147) Once it is agreed that language is inherently and inevitably opaque and inadequate, and that the expressions used in it are also inherently and inevitably fallible—a situation from which the techno-scientific discourse cannot escape—there is always room for improvement and enrichment in the form of critique. The improvement can be made with the help of translation, with an appreciation that the metaphorical nature of linguistic formulations invites a multiplicity of interpretations, a variety of translations, and as many as possible new alternative explanations. Whether "reality" remains stable or not, whether the rules of grammar change or remain fixed, whether the tradition of a speech community undergoes a major revolution or not—none of these factors alter the need to reconsider languages and their adequacy constantly and remorselessly.

Laor mentions Yehoshua Bar-Hillel's view of the "semi-open" language, one that allows translation from one language to another (in his example, between quantum theory and American-Indian Choctaw), because it poses the following conundrum. If every language is closed, the translation from one language to another is impossible. If every language is open, then it is possible to fully incorporate or integrate one language into another (as a form of radical reductionism). Yet, we find that languages do remain intact, that they do change over time, and that they are enriched by other languages. This situation leads Bar-Hillel to endorse the semi-open middle ground that promotes a rational communication of different speech communities across their linguistic boundaries. A similar concern with the openness of scientific discourses and practices has been the focus of debates among sociologists of science (e.g., Pickering 1990).

Furthermore, it remains an open question how much linguistic enrichment is possible through the process of translation, and to what extent such an enrichment either remains distant and tolerant à la Rorty or overtakes the very nature of a language. For instance, in the Spring of 1994, the French government and its Ministry of Culture decided to enforce a rigid set of rules limiting the number of English terms and expressions the French popular media could use, as if the infiltration of English into French would either demean French as a language or culture or undermine the national integrity of France. Incidentally, a nine-member Constitutional Council ruled later that

> while the use of French was compulsory [in broadcasting and advertising], the Government had no right to impose official French translations of foreign words on private citizens, companies and news media. It said the Government could only require their use by public authorities and public-sector companies other than radio and television. (Reuters, July 30, 1994)

The reference for the ruling was the 1978 Declaration of the Rights of Man and the Citizen, which is the preamble to the French Constitution.

Translation, then, is not limited to the inquiries of philosophers and linguists, semioticians and literary critics. The epistemological and ontological problems posed by the very notion of translation concern our perceptions of the reality that surrounds us, the ways in which we describe its various manifestations, the ways in which we communicate with each other, and the particular characteristics we uphold as social and political units. A brief digression into these issues pushed us into an intellectual trajectory that came full circle: I started with political claims and ended discussing the political concerns of governments about the necessary limits to the semi-openness of their (national and not merely natural) languages. But before I continue, I should perhaps point out what is political in the positions of both Quine and Chomsky, once again oversimplifying in the extreme.

Perhaps it would be useful to reintroduce Duhem, a precursor to both Quine and Chomsky who has pushed the national linguistic boundaries to an extreme both different from and similar to the one pushed by his nation eighty years later. Similar, because it brings up the boundaries between natural languages, and different, because he is more focused on science, while the contemporary French concern is with popular culture. According to Duhem, there is a difference between the mental dispositions of the French and the Germans: the former produce more intuitive scientists, the latter more rigorously deductive and experimental scientists. These ingrained differences are culturally bound and are taken to be the foundation on which science developed in Europe (Duhem 1991). That this view is chauvinistic and nationalist, that it sets up cultural fences only few can surmount, and that it falsely imprints human intellects and dispositions in particularized ways goes without saying. What is relevant for the present context is Duhem's extension of his conventionalist view of scientific theories to national, cultural, and political contexts.

Quine's view of the social nature of language can be construed as an egalitarian view of how people communicate. At the same time, if it is a social setting that eventually determines what linguistic rules ought to be followed and what linguistic expressions ought to be accepted as true or false, who exactly makes the decision? Does a certain linguistic leadership make up the rules? What is the relation between this leadership and the political leadership of the same community? Do representatives of each group negotiate the truth-value of all linguistic formulations? Is there an annual referendum or vote on the issues at hand? Is the decision-making process bound by a binary set of linguistic expressions that attempt to "describe" reality as opposed to "prescribe" moral standards? Besides, does the reality under consideration include the "political" reality of the state, as understood in the 1930s, for example, by the National Socialists in Germany? Quine's presumed egalitarianism may turn out to introduce a set of unintended political considerations and power relations.

Similarly, Chomsky's view of a deep structure lends itself to an egalitarian interpretation. If there is indeed a deep structure common to all languages, then presumably no matter how powerful the linguistic expression and its advocates in one language, it can be easily transported, translated, or transcribed into other languages more accessible to others. Moreover, every individual has the same opportunity to uncover the single deep structure, because it is "there." Everyone, then, has an equal opportunity to break the code and be the instrument for open communication among all speech communities. At the same time, one may argue that different people are positioned differently in terms of power relations, so that the pronouncements of some count for more than those of others. Moreover, the very proclamation that indeed there is a deep structure must be taken on faith, for unless such a structure becomes transparent to all, it is the word of one powerful group of linguists that carries the day. So Chomsky's alleged egalitarianism may turn out to have similar political problems to Quine's.

Let me recapitulate: language is opaque and so is the technoscientific discourse; different interpretations lead to different views of the world; and—as a central problem for the next century—these different views must be at least partially reconciled in order to allow for interdiscursive communication (call it reasonable or conventional). In allowing some sort of translation to be possible—that is, in allowing a form of connecting one discourse to another, however minimally, even as a useful critique—perhaps Lyotard's view of language games will be helpful.

Lyotard takes us back to Wittgenstein and the Vienna Circle, the roots of the debates in the twentieth century that have also informed Quine and Chomsky, and says:

> Wittgenstein's strength is that he did not opt for the positivism that was being developed by the Vienna Circle, but outlined in his investigation of language games a kind of legitimation not based on performativity. (Lyotard 1984, 41)

Wittgenstein set the tone for an exploration of the interaction between a multiplicity of language games, games whose rubric should not mislead anyone to believe that they are not taken seriously. The term *positivism* is being cast aside here, because it carries with it all the negative connotations one can attach to any theory or view with which one disagrees (just as Eco says the reverse about postmodernism). So what we have here is a concern with language games, their construction, their legitimation, and their potential for critical interconnection.

Legitimation crises have been the concern of those who, in one form or another, associate their intellectual efforts with those of the enlightenments (Ormiston & Sassower 1989, Ch. 3). Habermas has been the most vocal of those decrying the consequences of yet another legitimation crisis, one that would annul the progress of civilization, understood in emancipatory terms. Yet the quest for legitimacy, or at least for sound criteria according to which legitimacy is warranted, is itself problematic, for it may herald closure of discourses and the reduction of personal liberty rather than the reverse. After reviewing the three basic sources of legitimacy as they are expounded by sociologists—legitimacy as convention, as intellectual production, and as world coherence (Stanley 1978, 100-107)—Manfred Stanley has the following to say:

> Legitimation is an operation of closure. That is, it is an argument for discounting the value of pursuing further implications and for protecting established interpretations by means of enforced social sanctions. Legitimacy is always an unstable artifact of human interpretation—a dike against the never ending trickle, flow, or stream of scarcities. Because no legitimations are such as to be self-evidently applicable to all contingent situations, the interpretation of moral intelligibility is a task that falls to everyone. In that sense, every person is a moral agent, although some groups are always finding ways of concealing this fact from other groups. (Ibid, 131)

Stanley admonishes those concerned with legitimacy for three related reasons: first, because the quest for legitimacy is a quest for closure, for the abolishment of ongoing conversations or translations; second, because there is a false presumption concerning the stability of any legitimation; and third, because the moral agency of individuals may be denied in the face of contingent situations. I agree with this critique.

Though not always read within this particular context, Lyotard responds to the alleged crisis of legitimation of the twentieth century. To him, though, even if there is a crisis, it is unclear whether or not it should be rectified, reconciled, rebuilt, or repaired. His view, then, parallels Stanley's. According to Lyotard, "the game of science is thus put on par with the others," perhaps because "science plays its own game," or perhaps because "it is incapable of legitimating the other language games." (Lyotard 1984, 40)

In any event, Lyotard finds common ground with Quine's pragmatics, at least to the following extent: he believes that the use of languages is "subject to a condition we could call pragmatic: each must formulate its own rules and petition the addressee to accept them." (Ibid., 42) So indeed one is accountable to a speech community, a social group that, whether bound by tradition or national borders, accepts and uses the "rules" of a language, that is, its syntax. This goes further than the acceptance of the semantic rules according to which meanings are assigned to words and sentences in a language. And all of this brings us back to the question of politics. For in order to petition individuals and receive their assent, in order to find some social acceptance for the rules of the language game, each player must, according to Lyotard, "assume responsibility not only for the statements they propose, but also for the rules to which they submit those statements in order to render them acceptable." (Ibid., 62)

Again, each player must assume the responsibility (reminiscent of Haraway's plea mentioned in Chapter 2) of a player in a game. This is a requirement that transcends the common view of postmodernism as apolitical, for it refuses to accept the view that the "game" has been set in place in advance, and that its rules were determined prior to anyone making a move. The transcendence is possible especially within the context of what we imagine to be democratic institutions. Is it not the case that every move in and of itself legitimates the game, and that the adherence of players to its rules is a form of implicit acceptance of this legitimacy? If all of these questions are answered in the affirmative, then Lyotard is correct in beseeching us to play several games, "each of these games is interesting in itself insofar as the interesting thing is to play moves. And to play moves means precisely to develop ruses, to set the imagination to work." (Lyotard 1985, 61)

The issue for Lyotard is the empowerment of everyone who wishes to participate in any of the numerous language games available at any given time in particular places. To be fully engaged in language games is not only to assume the responsibility for their performances and proliferation (as Stanley advocates), but also to "figure out new moves. And even better . . . to invent new games." (Ibid.) Incidentally, this is quite different from the Habermasian view that reduces responsibility to a universal assent to rationality (the so-called Enlightenment Project) and to the achievement of a consensus among all participants (Habermas 1979), and different from Rorty's view of tolerance.

The invention of new moves is, of course, a difficult task, and the invention of new games an even harder undertaking. As Duhem already says in the context of laboratory work much earlier in this century:

> Every invention is a revolt: a revolt against the rules which it shatters because what they prescribe is false; a revolt against the methods from which it escapes because they show themselves incapable or mendacious; a revolt against the masters, whose over-narrow teaching it extends or whose false doctrine it overthrows. (Duhem 1991, 125–6)

Whether using the rhetoric of Duhem, Stanley, or Lyotard, one can imagine that it is possible to invent; one can bring about a revolt as one observes daily, in the workings of natural languages in popular culture, when rules are broken and new ones introduced, when linguistic traditions are challenged and overthrown. Linguistic games are unveiled routinely in rock concerts and video games, not only as artistic creations but also as means of recognizing a revolt, as Duhem calls them, or social critiques when petitioning addressees, as Lyotard calls them. Even when individuals recognize new moves for what they are and are seduced to enter new language games, how does that affect the political situation in which they live? How does that change their lives as moral agents, to use Stanley's term? Moreover, what is the relation between the new moves and the old ones, and between the new games and all the others that may still exist? We are back to ontology and translation.

As far as Lyotard is concerned, there is "a congruence between the domination of ontology and the torture inflicted to language games that are not those of ontological discourse." (Lyotard & Thébaud 1985, 53) It seems as if ontological discourses are set up at the pinnacle of a hierarchy or a pyramid of language games. That is, it is from an ontological basis that the rest of the language games receive their validity and legitimation. Any hierarchical setting is accompanied inadvertently by a mode of oppression, for the "higher" game takes precedence over a "lower" one, or, put differently, the lower the game on the totem pole, the more subordinate it and its rules are to those that are placed above them. In other words, any language game is always open to the "torture" of an ontological game because the latter can always undermine certain pronouncements and belittle, epistemologically and politically, whatever is said in the non-ontological game. Is there a way out of ontological domination?

Though concerned with the semiotics of television broadcasts and popular culture in general, Eco has this to say about the potential for freedom of the recipients of communication, those who in their capacity as audiences translate images and messages in their own ways:

> A herd of cows is perceived in the same way by an Italian and an Indian but for the former it signifies abundance of food, for the latter abundance of ritual occasions. So the suspicion grew that the sender organized the televisual message on the basis of his own codes, which coincided with those of the dominant ideology, while the addressees filled it with 'aberrant' meanings according to their particular cultural codes . . . these aberrations are seen as the last hope of freedom available to the defenceless masses. (Eco 1994, 90–91)

Eco's hope of freedom lies with the interpretive moves each recipient is allowed to make despite the intended or expected interpretations imposed by the producers and distributors of messages. Whether these messages are images or words, Eco's concern with the hegemonic oppression of addressees parallels that of Lyotard's.

One could argue that in order to avoid the oppression of one language game by another, all language games should be translatable, at least in principle, to a single ontological one so that their validity and credibility, their power to respond to criticism, would somehow be bolstered. But here Lyotard, just like Quine before him, is not quite sure that there is a uniform empirical foundation against which all games can and should be measured equally. That is, even if the same set of sense data produce competing theories and games, it is not clear that each is reducible to every other. At this juncture, Lyotard adds another wrinkle to any examination, for he distinguishes between languages and language games. In his words:

> It so happens that languages are translatable, otherwise they are not languages; but language games are not translatable, because, if they were, they would not be language games. It is as if one wanted to translate the rules and strategies of chess into those of checkers. (Ibid., 53)

It seems, then, that language games are different from languages, so merely to substitute the former for the latter will not do. If languages are more broadly construed as natural languages—those within which translations are possible—and language games are construed in more narrowly defined terms as those circumscribed discourses whose rules are rigid and internally particularized, then it would help to distinguish between the two categories. But still there is a great deal of overlap, for languages have (syntactical) rules as well, those of grammar. And just as it is impossible to translate the rules of chess to those of checkers, it would seem impossible to translate the grammatical rules of one language to those of all others. (Unless, of course, one reverts to the Chomskian sense of one deep structure underlying all rules.) We are back to the continuum of positions that lie between Quine, the social conventionalist, and Chomsky, the foundationalist (or essentialist). We are also back to the political backdrop against which this chapter was constructed, for we can return now to the different schools of thought I have enumerated earlier and inquire whether their discourses constitute languages in Quine's sense or language games in Lyotard's sense.

The Politics of Games

There is, though, a nagging question that haunts the critics of postmodernism in general, a question whose urgency overshadows my concern with the difference between languages and language games: what about responsibility? Though Christopher Norris is more concerned to defend Habermas's critique of Derrida, his insistence on "the various *specific* normative dimensions that exist within the range of communicative action" (Norris 1990, 58) is relevant here, for it shifts the discussion from general epistemological concerns to moral ones.

Responsibility is contextualized in terms of moral agency, of the author or the reader, and is understood by critics of postmodernism (whether attacking Derrida or Lyotard or any of their colleagues) in terms of the "demise of the author and of causality." As Pauline Marie Rosenau says:

> Because no single human being can be held accountable for a situation in the sense of having causal input, no one "authors" a text-event as such. For the social sciences the death of the author results in removing responsibility from human subjects. (Rosenau 1992, 33)

Rosenau maintains that it is the loss of causality in the performances of textual analyses, their production, distribution, and consumption, that brings about the denial of any personal responsibility (Ibid., 56). The lack of personal responsibility is a consequence of the odd position into which postmodernists have put themselves, a position that is tied with relativism. In her words:

> If post-modernists were consistent and true to the linguistic relativism they expound, and if they really believed successful communication was impossible, then they would cease to deconstruct what others wrote or said. To avoid being hypocritical, they would have to remain silent . . . If post-modernists were sincere when they stated that no statements are privileged, not even their own, then they would have to give up their right to speak without authority. Only this would assure internal consistency. (Ibid., 178)

First, to what conditions of consistency is Rosenau appealing here? Second, is it useful for her purposes to add *ad hominem* terms such as "hypocritical" and "sincere"? Third, does she mean "with" or "without" authority when she speaks of postmodernists speaking? Fourth, is silence the only way to respond to a multiplicity of interpretations? Why not have a dialogue, something like a Platonic or Hegelian dialectic? Fifth, what is successful communication? And do postmodernists indeed believe it to be impossible? These are all rhetorical questions, of course, for I find the tone and indictment of this passage intellectually insulting. There is no reason for public humiliation if there is room for error on the part of the observer or critic.

Yet, the spirit of Rosenau's argument reflects the frustration of many critics of postmodernism, and therefore it deserves attention. It exemplifies a whole range of critiques that assume a definite lack of political commitment by postmodernists, while overlooking how similar some of the concerns of postmodernists are to the position of so-called analytic philosophers of language or social theorists. It is this view that I wish to repudiate in this book, for I believe that there is a strong political commitment on the part of most postmodernists, though it is a commitment that is announced neither in traditional vocabularies nor within classical institutions. I have already mentioned and shall return to this political theme in every chapter of this book. Back to language games.

I appealed earlier to Wittgenstein's sense of language games in order not to rely exclusively on Lyotard's sense of these games and the presumptions of frivolity that haunt all postmodernists, their discourses and pronouncements. Yet, there may be something appealing about retaining a more playful character in relation to language games, while retaining the seriousness of languages in general. There is no reason to dismiss the significance of language as such, for without it communication would be impossible, and without communication it would be impossible to conceive of democratic institutions. The different uses of language eschews the trap of the cliche "might makes right," for they are essential tools with which to weave a social fabric and challenge power relations. What about language games?

Given Lyotard's terminology, language games can be understood as derivatives, second-order phenomena, whose very existence depends on some level of communication already in place within natural languages. Yet these games are my focus in this book, as the next two chapters will illustrate. One case is that of Popperians and some postmodernists; the other is the case of some feminists and Popperians. Unlike the concerns of translation from one natural language to another, or from one natural language to an artificial one (of logic or computer software), I try to explore how discursive games of schools of thought within the same language can confront or be made to engage each other. Perhaps my illustrations will fail the test of a translation in a Quinean sense, but they will at least demonstrate how I understand the Lyotardian, postmodern sense of translation and intellectual open-mindedness.

The success or failure of a postmodern translation in attaining political goals of destablizing technoscientific discourses can be measured in terms of the following criterion: can the translation find common critical features and techniques of challenge without thereby obliterating differences? So the mode of translation and its attendant results will not bring about an epistemological alliance that in the name of a political goal overlooks inherent differences and contradictions. There is always a suspended judgment about truth, objectivity, reality, or what have you, because multiple perspectives must be accounted for before that judgment can be even ascertained and evaluated. And making a suspended judgment, no matter how temporal, does not avoid assuming a great deal of personal risk and responsibility.

Chapter 4

Postmodern Technoscience: A Critical Engagement

Introduction

Most intellectuals keep postmodernism and philosophy of science at arm's length, as if the two have little in common except the academic appointments of their champions. In this chapter I critically examine two related claims: first, that postmodernism and philosophy of science depend on each other in a manner similar to the Enlightenment and Romanticism, that is, they respond to and dispute each other's claims. I mean this neither as a claim about a reaction from the one to the other, nor in the sense used by David Bloor in reference to Mannheim's characterization of the difference between the Romantic and Enlightenment ideologies (Bloor 1991, 62–76). And second, that what underlies and emanates from both postmodernism and philosophy of science is a political perspective and commitment. These claims suggest not only the possibility of translating from one area to the other when they are critically engaged with each other, but also the potential for using both areas simultaneously in order to transform and possibly improve the human condition. In this respect, then, this chapter (unlike Chapter 2, which also examines postmodern technoscience) exemplifies my sense of a translation from one language game or discourse to another. Let me explain.

Lyotard claims, as we have seen in the previous chapter, that one cannot translate the rules of chess to those of checkers. Yet, the incommensurability between these two games is not extreme, for both games are played on the same board, an eight by eight grid. In many cases, this fact alone induces manufacturers to market them in one box. It is true that most moves are different, but there are also similarities between setting the two players across from one another with an expectation to cross the board to the opponent's side (For more on this, see Wittgenstein 1968, 31–32). I do not wish to extend this example too far; only far enough to explain the problematic task I set for myself in this and the following chapters. Though chess and checkers are and will remain different games, it may be fruitful to transfer the strategic orientation of one into the other: for instance, plan ahead a series of moves while accounting for possible countermoves and do not limit your vision to the immediate next move as a quick reaction to your opponent's move.

Modernist Philosophy of Science and Romantic Postmodernism

As far as the literature goes, there is a fundamental rift between what is understood as philosophy of science and postmodernism, as already explored in Chapter 2. The rift can be captured by a variety of binary conceptual references: objectivity and subjectivity, truth/certainty and relativism, rational knowledge and superstition, or modernism and postmodernism. To a certain extent, it seems that the rift is so polarized that it can never be bridged. This view of the situation has entrenched an attitude among intellectuals, especially in the academy, so that one must adhere to only one of these mutually exclusive conceptual matrices. The adherence to one of these conceptual matrices has turned an epistemological choice into political loyalty.

It seems that the received view described above is politically dangerous and epistemologically misguided. I do not merely suggest that philosophers of science should encounter and critically engage postmodernists, for such encounters can be reconstructed even in the works of Bacon (if he is understood simultaneously as an inductivist and the forbearer of modern science and as an interpreter of natural phenomena in the broadest sense of interpretation of empirical observation). For example, Bacon's first aphorism in the *New Organon* reads:

> Man, being the servant and interpreter of Nature, can do and understand so much and so much only as he has observed in fact or in thought of the course of nature. Beyond this he neither knows anything nor can do anything. (Bacon 1985, 39)

Instead, it seems to me that perpetuating the conception of the incompatibility or incommensurability of the ideas and views of philosophers of science and postmodernists is counterproductive for the improvement of the human condition. In this respect, then, I want to amplify the potential for translation among intellectuals who see themselves as political rivals.

I should hasten to add a few disclaimers. Philosophers of science do not constitute a uniform group of individuals who agree with each other about the theoretical foundations and practical application of science. Besides, philosophers of science include both philosophers interested in science and scientists who are philosophically minded. Moreover, what constitutes science for all of them is an open question that has not been settled by demarcation criteria proposed, for example, by the Vienna Circle or Popper. Furthermore, the differences between the social and the natural sciences are still a bone of contention among them, just as the categories of mature science and preparadigmatic science are not clearly defined or ascertained. In addition, science itself is not a uniform enterprise even among those who agree about the delineation between it and pseudo science or between social and natural science, but is perceived as a complex of views and practices that at times illustrates clashes and crises of faith in the entire enterprise, what Kuhn calls paradigm shifts (1970).

Like their philosophical counterparts, postmodernists have as many shared beliefs as they have radical misgivings about their colleagues' works. So, to speak of postmodernism as if it were a monolithic edifice is a misleading oversimplification. And finally, the transformation and improvement of the human condition is a problematic goal because of its implicit paternalism and teleology as well as its explicit belief in some form of progress, however openly defined. Perhaps by the end of this book some of these concerns will be laid to rest, perhaps others will be brought to light. I plead for some latitude needed in this particular context for an examination of an enormous terrain whose boundaries defy clear delineations.

I must also insist that this critical reexamination need not result in a definite alliance between the alleged rival factions; yet, it may sharpen a critical appreciation of the similarities and differences of these two groups. I am far from advocating a Habermas-like roundtable at which warring parties negotiate away their differences; nor am I advocating a peaceful or tolerant coexistence (Rorty's) whereby each party is encouraged to ignore the other. Instead, I submit that a critical engagement between these two parties who seem to be concerned with the improvement of the human condition is more reasonable than either bluffing around a poker table or a continued feud in which the contestants—for power, prestige, or money—speak at cross-purposes. Does this sound too naive or even romantic? Perhaps.

Scientific Communication

The romantic spirit is very much with us today. I call it a spirit because I do not want to associate it with a canonical, historical period that accords with English literary genres or art history categories. I call it a spirit or an orientation because of its emphasis on individual aspirations and dreams. The romanticism I have in mind is the one that heralds subjectivity and validates personal perceptions and sensations, the one that in Nietzschean terms privileges the one over the herd mentality of the masses. This orientation is attributable to postmodernists who draw heavily on the Nietzschean legacy and insist on multiple readings of any text.

At the same time, the enlightenment spirit is with us as well, the spirit that appeals to the objective, the value-neutral reality with which we must come to terms; an unforgiving reality that supersedes the personal aspirations of individuals. This orientation is attributable to scientists and philosophers of science who attempt to overcome the contingencies of personal testimonies and insist in their stead on intersubjective reports that can be validated by independent sources (e.g., repeatable experiments or the standards of the Royal Society).

Romantics and postmodernists, following the classification offered above, legitimate the perspective of the individual, while Enlightenment-modernist philosophers of science legitimate the objectivity that supersedes the arbitrariness of individual perspectives. Each group views its position as fundamental, if not foundational. And both positions belie a certain conundrum: an inherent problem of communication about one's surrounding. From one perspective there are only different individual perspectives so a shared grounding is missing, while from the other perspective all one has is a shared grounding without any individual anchoring.

At times this problem is formulated in terms of relativism, as we have seen in previous chapters when critical comments are made regarding postmodernism. For Ernest Gellner, the danger of relativism is its logical (not psychological) consequent nihilism (Gellner 1992, 54), a cognitive (and not moral) nihilism that is epistemologically false (Ibid., 71) and which undermines the possibility and eventual stability of communities. Gellner, as a Popperian anthropologist, minces no words in his condemnation of postmodernism:

> Postmodernism is a movement which, in addition to contingent flaws—obscurity, pretentiousness, faddiness, showmanship, cultural name dropping—commits major errors in the method it recommends: its penchant for relativism and preferential attention to semantic idiosyncrasy blind it to the immensely important, absolutely pervasive asymmetry in cognitive and economic power in the world situation. (Ibid. 70)

I am interested in disputing not Gellner's reading of the concern of postmodernism with economic power (e.g., Harvey 1989), but with his linkage of relativism (Romantic or not) with postmodernism, a theme that permeates the discomfort felt in his book in relation to the political significance of this "movement."

Postmodernists and philosophers of science are aware of their own positions and the need to overcome the obstacles for (scientific and political) communication with each other, especially if relativism is the ghost that haunts them both. Whether one's point of departure is Archimedean or personal, the journey, even if not on the same "line," can be along parallel lines that acknowledge each other's existence and usefulness. Such a recognition may require a translation from one line (of thought) to another, a translation that is at most temporal and provisional, but can be both critical and informative. This situation has been discussed in previous chapters, but its presentation here is from another perspective. Put differently, the scientific enterprise, oddly enough, parallels postmodernism: with hegemonic pretensions intact, these enterprises still contain confusion and critical self-reflection.

The models and theories of scientists, as Popper claims, are putative: they change whenever there are empirical reasons to change them (Popper 1963). When a change becomes ominous, when it can no longer be avoided, an entire edifice is replaced with a more accurate one (that is, one that approximates reality more closely given the available data). This situation illustrates the instability of scientific knowledge-claims and the need to be prepared for unknown possibilities. This situation may lead to despair or hope: despair regarding the futility of one's work, knowing that whatever is taken for granted epistemologically on one day may be challenged and undermined on another; and hope regarding the continued efforts to replace inaccuracies and mistakes with truths. A similar attitude permeates the work of postmodernists who celebrate ambiguities and novelties, providing spaces for a multiplicity of interpretations without regard to the eventual demise of knowledge-claims or cherished truths.

Lyotard's Postmodern Philosophy of Science

Is there already a slight recognition that the conjunction of postmodernism and philosophy of science or technoscience is possible? It may be helpful to enumerate some features or characteristics associated with postmodernism in order to evaluate the possibility of its conjunction with or relevance to (philosophy of) science. As noted in Chapter 2, it is difficult and futile nowadays to come by a simple definition of deconstruction, poststructuralism, or postmodernism. To some extent this is because of the problematics of definitions, whether one falls into the trap of essentialism or remains in the nominalist or instrumentalist fold. Postmodernism defies definition for another reason as well: the very activities of postmodernists have to do with proliferating a multiplicity of definitions, modes of linguistic analysis, and experimental narratives. In this light, then, to uphold any unique label would be at once mistaken and misleading, as if one could capture the flavor of postmodern practice with a definite instrument of language and thought. Incidentally, these comments are equally applicable to Romanticism.

Despite the inability and unwillingness of many to define postmodern practice, and regardless of my attempts to do so in previous chapters, I would like to follow Lyotard in providing a working definition, a wedge with which to begin the discussion of how postmodern philosophy of science or postmodern technoscience may differ from other attempts to organize and explain a technoscientific discourse and enterprise. An examination of the epistemological themes present in the writings of some postmodernists will also politicize the discussion of technoscience.

Just as historians of science classify the Scientific Revolution with the onset of modernity (e.g., Cohen 1985), critics of postmodernism (e.g., Callinicos 1990) historicize and periodize postmodernism. In following the historians' lead, these critics ignore Lyotard's warning that what is at issue is a "postmodern condition," and not a specific time in history (Lyotard & Thèbaud 1985, 16n). For example, when I explore the epistemological possibilities and political opportunities afforded to contemporary society in postmodern terms, I am not only attaching my discussion to certain postwar, French philosophers, but also to certain comments (quoted earlier) already made by Bacon about the notion of interpretation that accompanies the collection of empirical data. At least three themes can be elucidated from the concerns of postmodernists with scientific knowledge.

One common theme that permeates postmodernist writings on scientific knowledge is its relation to all other forms of knowledge, and the examination of knowledge in terms of language, as examined in Chapter 2. According to Michel Foucault, (classical notions of) knowledge "consisted in relating one form of language to another form of language; in restoring the great, unbroken plain of words and things; in making everything speak . . . The function proper to knowledge is not seeing or demonstrating; it is interpreting" (Foucault 1970, 40). Interpreting, as already noted by Bacon, is linked to a notion of organization or order: "What makes the totality of the Classical *epistème* possible is primarily the relation to a knowledge of order" (Ibid., 72). Incidentally, this view of classical episteme recalls the modernist notions of the Enlightenment ideals of scientific knowledge.

The importance of languages in Foucauldian terms, or language games (understood as frameworks and models) in Lyotardian terms, is their presumed transparency for the purpose of studying the history of ideas, the development of science, and the changes that civilization has undergone over the years. In Foucault's words:

> Languages, though imperfect knowledge themselves, are the faithful memory of the progress of knowledge towards perfection. They lead into error, but they record what has been learned. In their chaotic order, they give rise to false ideas; but true ideas leave in them the indelible mark of an order that chance on its own could never have created. What civilizations and peoples leave us as the monuments of their thought is not so much their texts as their vocabularies, their syntaxes, the sound of their languages rather than the words they spoke; not so much their discourse as the element that made it possible, the discursivity of their language. (Ibid., 87)

Knowledge is linked to language, language bespeaks order, and order underlies the conception of science—as a regulating ideal or as a principle of operation. Following Foucault and Lyotard, one may ask: is all knowledge nothing but scientific knowledge? Is scientific knowledge the "core" in Imre Lakatos's sense of the scientific enterprise of "research programmes," and other knowledge claims part of its "protective belt" (Lakatos 1970), open to continued change?

According to Lyotard, "scientific knowledge does not represent the totality of knowledge; it has always existed in addition to, and in competition and conflict with, another kind of knowledge, which I will call narrative in the interests of simplicity" (Lyotard 1984, 7). Moreover, once scientific knowledge is understood in terms of other kinds of knowledge—that is, not simply as the basis on which all other forms of knowledge are constructed—it is possible for Lyotard to claim that "The game of science is thus put on par with the others" (Ibid., 40). Once there is a multiplicity of language games, what distinguishes their different use? As noted in the previous chapter, "These languages are not employed haphazardly, however. Their use is subject to a condition we could call pragmatic: each must formulate its own rules and petition the addressee to accept them" (Ibid., 42). In this respect, there is an agreement between postmodernists and philosophers of science regarding the pragmatic contextualization of their discourses, at least in terms of the rules according to which the discourses are used for either explanatory or predictive purposes.

Another theme characteristic of postmodern views of science has to do with the distinction between "modernism" and "postmodernism" (despite Latour's protestations, as noted in Chapter 2). While modernist views are commonly traced to the Enlightenment Project (one that attempted to provide the foundations of all possible knowledge), postmodernist views reject the possibility of such an undertaking and limit their reconstructions to the provision of a multiplicity of epistemological models. According to Lyotard, "Simplifying to the extreme, I define *postmodern* as incredulity toward metanarratives" (Ibid., xxiv).

Though Foucault juxtaposes classical knowledge of order with a more fragmented kind (Renaissance) that is associated with the "sciences of man" and later with the "social sciences," it is possible to distill from his distinction a relation between so-called modernists and postmodernist views of knowledge (not unlike the problematic distinction between the Enlightenment and Romanticism alluded to earlier). As for classical knowledge, he considers it to have been the positivist dream:

> Language was a form of knowing and knowing was automatically discourse. Thus, language occupied a fundamental situation in relation to all knowledge: it was only by the medium of language that the things in the world could be known . . . It was in language that all generality was formed . . . This is the positivist dream of a language keeping strictly to the level of what is known. (Foucault 1970, 295–296)

The other kinds of knowledge, by contrast, broke the canonized notion of order through "the fragmentation of language" (Ibid., 307). If the fragmentation is carried to its logical extreme, then personal accounts become the standard of linguistic communication in contrast to the standard of an ordered language game with a clear ontological foundation in the positivist sense. Is it possible, as Wittgenstein asks, to construe personal accounts as "private languages" or are they (through their use) inherently mediated and therefore public, conventionally speaking? (Wittgenstein 1968)

Unlike modernists who are committed to providing metanarratives, that is, comprehensive metaphysical frameworks within which all knowledge claims can be organized, tested, and explained, the postmodern "condition" prevents this untenable pretense without, presumably, reverting to the incomprehensibility of private language games. Once metanarratives no longer provide the backdrop against which to measure everything, the reductionist need—the need to reduce one set of propositions to another, more fundamental and thereby more secure—is mitigated, if not completely eliminated. In Lyotard's terms:

> Postmodern knowledge is not simply a tool of the authorities; it refines our sensitivity to differences and reinforces our ability to tolerate the incommensurable. Its principle is not the expert's homology, but the inventor's paralogy. (Lyotard 1984, xxv)

The tension between the two extreme views delineated above in terms of romantics and modernists is recognized by postmodernists interested in science. Their attempt, then, is to maintain a balancing act or dialectical tension between two untenable positions into which one may fall when adopting either of the two views of knowledge.

A third theme in postmodern views of science connects knowledge to power within the context of the social, political, and economic conditions of capitalist society. In Lyotard's words: "Knowledge . . . is . . . a major—perhaps *the* major—stake in the worldwide competition for power." (Ibid., 5) But what kind of power is at stake? Is it only the power of capitalists, for example, to exploit the fruits of technoscientific knowledge so as to increase their profits?

Marx and Engels connect the industrialist powers of commerce to the foundation and practices of natural science, so that industrial practice is not merely a question of exploiting the fruits of science but a question of what motivates and guarantees the activities of science. Discarding Feuerbach's idealist view of scientific research, they claim:

> Feuerbach speaks in particular of the perception of natural science; he mentions secrets which are disclosed only to the eye of the physicist and chemist; but where would natural science be without industry and commerce? Even this "pure" natural science is provided with an aim, as with its material, only through trade and industry, through the sensuous activity of men. (Marx and Engels 1988, 63)

Lyotard continues along these Marxist lines and stresses that the notion of power in relation to scientific knowledge is pervasive: "Knowledge is no longer the subject, but in the service of the subject: its only legitimacy (though it is formidable) is the fact that it allows morality to become reality." (Lyotard 1984, 36)

Power, according to Lyotard, is not only the consequence of the mastery of knowledge, as some would have us believe in light of Bacon's aphorism "human knowledge and human power meet in one" (Bacon 1985, 39); it also "legitimates science and the law on the basis of their efficiency, and legitimates this efficiency on the basis of science and law. It is self-legitimating, in the same way a system organized around performance maximization seems to be" (Lyotard 1984, 47). As such, then, knowledge production, dissemination, and consumption are self-validating processes that empower themselves as they unfold (Ibid., 54). I will return to the theme of power and science when examining the inherent politicization of postmodern philosophy of science.

Toulmin's Postmodern Philosophy of Science

Lyotard's *The Postmodern Condition* was originally published in French in 1979, then translated and published in English in 1984. From what I have been able to ascertain, the first use of the term "Post-Modern

Science" in 1981 (and later in 1985) is attributable to Toulmin. Toulmin denotes post-modern science with an emphasis on the hyphen; that is, for him the classification (following, as he says, Frederick Ferre) is exclusively historical and not merely historically informed as one would gather from Lyotard's disclaimer. As mentioned before, for Toulmin there has been a "dual change" in the posture of scientists toward their subject matter and in the "world picture" they come up with. Modern scientists viewed themselves as "spectators," while post-modern scientists recognize that they are in fact "participants" in the study of nature; modern nature was considered a "self-contained, deterministic mechanism," while post-modern nature is less mechanistic and detached so that it "reintegrate[s] humanity with nature" (Toulmin 1981, 71).

In demarcating between modern and post-modern science, Toulmin focuses on four main features that have been transformed by the twentieth century: first, the separation between scientific thought and technological practice; second, the philosophical foundations that underlie scientific theories; third, the presumed connection between objectivity and value-neutrality; and fourth, the professionalization of scientific activities (Ibid.). Toulmin's demarcation parallels the ones offered above, such as between the Enlightenment and Romanticism or Foucault's Classical and Renaissance view of knowledge. I would like to add that my report of these demarcations is in no way an endorsement, for I have many misgivings about their accuracy and efficacy. Nonetheless, it may help here to mention just one epistemological change of heart in order to appreciate the sort of historical reconstruction undertaken by Toulmin.

The modern view of scientists as "Omniscient Calculators" (Laplace's term)—outside observers who can fully apprehend the intricacies of natural phenomena and figure out how to explain and predict them—has been replaced, according to Toulmin, with the post-modern view of scientists as participants in the study of natural phenomena. It is interesting to note here how differently Toulmin and Forman document these changes. Toulmin strings together Einstein's "dissatisfaction with late nineteenth-century attempts to link Isaac Newton's system of mechanics with James Clerk Maxwell's theory of electromagnetism" and the critiques of Werner Heisenberg and Niels Bohr so as to explain the foundational changes required by relativity and quantum mechanics. "For the purposes of physics, any one viewpoint was as good as any other . . . the relativistic principle implied that any frame of reference was physically as good as any other." (Ibid., 92–93) So twentieth century science no longer presumes that scientific measurements are objective in some universal sense, or unrelated to the individuals who perform them.

Perhaps the notion of interpretation is not fully emphasized by Toulmin, as in the case of computer-generated images that are supposed to represent micro- and sub-atomic phenomena (more on this in Chapter 7), but he is correct in stressing the impossibility of claiming any validity and certainty for the empirical tests undertaken by scientists. The best scientists can do is explain what particular parameters they have chosen and why they have chosen them, and then articulate whatever they wish within this limited framework. Their posture as interpreters is similar to the one advocated for them by Bacon, Lyotard, and Forman.

If, as Toulmin continues to explain, "all scientific understanding whatever involves uneliminable intervention by the scientist in the processes that he is seeking to understand" (Ibid., 96), then the view of post-modern scientists as participants makes sense. Their view of "nature" and of their work must take into account the possibility of integrating their social and political concerns with the models they construct as scientists. Following J. D. Bernal and Michael Polanyi, Toulmin acknowledges that the sort of neutrality asserted by modern science has become untenable for post-modern science. In his words: "After Hiroshima, however, the monastic attitudes toward the work of science have become increasingly difficult to sustain." (Ibid., 112)

Here at the end of the twentieth century, as Toulmin agrees, in order to speak of postmodern technoscience, it is necessary to consider social and ethical issues, to figure out the political and economic frameworks of technoscience (e.g., the SSC), and the internal traditions and power relations of the technoscientific community.

Feminist Postmodern (Philosophy of) Science

My account would be incomplete if I did not reincorporate feminist critiques of science into my translation, not only because of the role they have played in challenging the hegemony of science, but primarily because of their contributions to the politicization of science and its philosophical scaffolding. At the same time, I have already acknowledged the problematic attempt to conjoin postmodernism and philosophy of science, so adding to them feminist theory and critique would seem almost impossible. Hence my all-too-brief comments.

According to Sandra Harding, postmodern feminism challenges the assumptions of feminist empiricism and feminist standpoint epistemology—the former closely linked to some modernist or positivist views of science, the latter much more concerned with contextualizing knowledge-claims without complete loss of the notion of objectivity (Harding 1986, 26). Unlike both of these feminist models (which are aligned with the goals of modern science concerning the anchoring of knowledge claims in some objective, observable reality), postmodern-feminist views of science undermine any such claims for universality. Moreover, the sort of universality and objectivity to which these two models aspire turns both models into contenders for what has been understood by feminists to be a "successor science" to the male-dominated, sexist, racist, and classist science that has evolved since Bacon (Ibid., 142).

The tension that becomes apparent when postmodernism is linked to feminist conceptions of science can be understood in the following terms: both are critiques of modern science, yet their proposals for change differ because of different political agendas. Jane Flax (quoted by Harding) describes "the successor science and postmodern tendencies in feminist epistemology as conflicting." (Ibid., 151) The issue is politics: either feminists fight to undermine male-dominated science with a specific political agenda that replaces not only the language of science but its practices as well, or they will undermine their own agenda by accepting the political neutrality that seems to underlie postmodern refusal of foundations and universality. At another point (Ibid., 154), Flax is quoted to the effect that there is an "affinity" between postmodernist and feminist critiques.

Perhaps one way to appreciate this notion of affinity is to focus on the insistence, by Harding as well as by Lyotard, that incoherent views may be simultaneously proposed, that one may not find absolute answers to difficult questions about the nature of physical phenomena, that a foundation is pertinent even if temporal, and that the political dimensions of scientific inquiry are crucial ingredients for the understanding of the results offered by scientists (Ibid., 164).

Unlike Harding, Haraway responds to the tension between some postmodernist and feminist views by setting the stage of the debate in her own way. In her scenario, feminist empiricists and postmodernists are the interlocutors who need to change "metaphors" and think of situated knowledge as a way out of the tension or dilemma (Haraway 1991, 188). The dialectical shift is not from empiricists and situated-knowledge proponents to postmodernists, but rather from empiricists and postmodernists to situated-knowledge proponents. According to Haraway, "the feminist standpoint theorists' goal of an epistemology and politics of engaged, accountable positioning remains eminently potent. The goal is better accounts of the world, that is, 'science'." (Ibid., 196)

What situated-knowledge claims bring to the study of science is an inherent combination of epistemology and politics; they are unable and unwilling to claim value-neutrality for scientific research. Haraway's juxtaposition is similar in tone to those I have already alluded to in the conception of postmodern philosophy of science, even the historically defined juxtaposition of Toulmin. If scientists are perceived socially and politically, and if their work is contextualized culturally, then the modernist notions of the Enlightenment disappear (as Toulmin insists). Haraway echoes Toulmin's description of postmodern scientific inquiry when she says:

> Situated knowledges require that the object of knowledge be pictured as an actor and agent, not a screen or a ground or a resource, never finally as slave to the master that closes off the dialectic in his unique agency and authorship of "objective" knowledge. (Ibid., 198)

Toulmin speaks of scientists as participants, and Haraway speaks of them as actors and agents. It is true that Bacon speaks of them as servants, but he also speaks of them as interpreters; and as interpreters, they become actively engaged in their observations. If there is a common thread, no matter how thin or thinly disguised, it is a feminist and postmodern recognition of the need to contextualize science to the extent that people do matter and their identities do make a difference in the construction and acceptance of scientific ideas. As such, epistemology cannot be devoid of politics.

The Politics of Postmodern Technoscience

As mentioned in Chapter 2, there are some general objections to the spirit of postmodernity as described so far, objections that delegitimate the possibility of postmodernism contributing anything (epistemological as well as political) to ongoing debates about science. I return to several of these objections in order to assist myself in my role as a semi-open and semi-fair translator. A one-sided polemic for or against a postmodern infusion into the technoscientific discourse would undermine the possibility of a translation altogether.

The main objection to postmodernism is its claim to be "post" modernist (e.g., in Toulmin's sense); it is in fact not different from the modernist project itself, albeit with some criticism and revisions. Moreover, postmodernism sheds no new light on similar, and much earlier, debates undertaken by Enlightenment leaders and their Romantic critics (Callinicos 1990, Ch. 1). For that matter, as seen in Chapter 2, concerns about relativism and skepticism are apparent within the citadels of analytic philosophy, the presumed bastion of scientific apologetics in Locke's sense of the role of the philosopher.

Less concerned with the question of "originality" than with the interplay of the notions of action/reaction (in terms of the motivation for a response on a cultural level), Felix Guattari aptly remarks:

> I am not taking the easy way out to demonstrate that postmodernism is nothing more than the death throes of modernism: it is a reaction, and in a way, a mirror, of the formalist abuses and simplifications of modernity,

> which has not in fact succeeded in differentiating itself from that movement. (Guattari 1986, 40)

In addition, many critics have become tired of hearing of yet another "crisis" in the activities and perceptions of intellectuals. Is history an amalgam of series of crises? Lyotard acknowledges this point when he says:

> Are "we" not telling, whether bitterly or gladly, the great narrative of the end of great narratives? For thought to remain modern, doesn't it suffice that it think in terms of the end of some history? Or, is postmodernity the pastime of an old man who scrounges in the garbage-heap of finality looking for leftovers, who brandishes unconsciousnesses, and who turns this into the glory of his novelty, into his promise of change? But this too is a goal for a certain humanity. A Genre. (A bad parody of Nietzsche. Why?) (Lyotard 1988a, 136)

Speaking of postmodernism as a "garbage-heap of finality" is not a flattering way to describe one's thought and action. Yet it is a way of being honest about one's sense of being an "old man" in the sense of having learned quite a bit of history, and of being engaged in a "pastime" in the Hegelian sense of the owl of Minerva whose insights are at best retrospective. But of course one need not take Guattari's and Lyotard's pronouncements as negative indictments; they may be invocations and pleas for legitimacy, despite the peculiar shades and hues with which they are produced, distributed, and consumed.

Added to the concerns about originality and the definition of a period in misleading terms—namely, that there is no such a thing as postmodernism, either modernism or nihilism—there is also a concern over the specific answers and judgments that postmodernism will be able to generate. As this book tries to argue, the particular brand of politics that can be associated with postmodernism is quite baffling for those accustomed to political frameworks such as Weber's, Marx's, or Habermas's. Is the call for linguistic—and therefore metaphysical and political—multiplicity not an abdication of responsibility and an invitation for complete chaos and anarchy of the worst kind?

Once scientific discourses as narratives are multiplied, following Lyotard,

> There thus arises an idea of perspective that is not far removed, at least in this respect, from the idea of language games. What we have here is a process of delegitimation fueled by the demand for legitimation itself. The "crisis" of scientific knowledge, signs of which have been accumulating since the end of the nineteenth century, is not born of a chance proliferation of sciences, itself an effect of progress in technology and

> the expansion of capitalism. It represents, rather, an internal erosion of the legitimacy principle of knowledge. There is erosion at work inside the speculative game, and by loosening the weave of the encyclopedic net in which each science was to find its place, it eventually sets them free. (Lyotard 1984, 39)

Lyotard appears to concede the legitimation crisis announced by the Frankfurt School, as mentioned in Chapter 3. Yet he does not follow its particular mode of critique. For instance, on the notion of consensus, Lyotard says:

> The network formed by all these phrases, for which no common code exists, becomes more fragile in proportion to its increasing complexity. It seems to me that the only consensus we ought to be worrying about is one that would encourage this heterogeneity or "dissensus." (Lyotard 1988b, 44)

Moreover, once the Habermasian ideal of consensus (which presumes that people will agree to use the same rules and that consensus is the appropriate goal) is discarded, Lyotard continues, there is "a recognition of the heteromorphous nature of language games," which in turn "sketches the outline of a politics that would respect both the desire for justice and the desire for the unknown" (Lyotard 1984, 66–67). The Frankfurt School would join in the fight for justice, but would its disciples respect the desire for the unknown? Given Habermas's agenda and his encyclopedic efforts to master and classify all knowledge-claims, I doubt it.

The connection between epistemology and politics is inherent in the view of a multiplicity of language games for one can no longer assume the indisputable validity of a canon, be it a scientific model, method, or theoretical framework (especially, as feminists argue so well, when the canon is sexist, racist, and classist). "This implies," according to Lyotard, "that the task is one of multiplying and refining language games." (Lyotard & Thébaud 1985, 49) And this, of course, should not be confused with a recommendation to provide a new foundation. As he explains:

> I mean that, ultimately, what does this thesis lead to? To a literature, in the best sense of the term, as an enterprise of experimentation on language games, to a general literature if one can put it this way. (Ibid.)

One could object to Lyotard's use of the concept of experimentation, for it connotes something quite definite for scientists working in the laboratory. However, the postmodern sense of experimentation is not that dissimilar from modernist views of science: an experiment may yield unintended results; it may refute previously accepted assumptions; and it may lead to results whose relevance may be missed for generations. There is some free play in experimentation: not all experiments need be designed and undertaken within the Kuhnian (paradigm) prison that limits everyone to solve prefigured puzzles.

The turn toward a multiplicity of games in science (not intending any disrespect by calling them games), what Feyerabend calls pluralism or anarchism of different applicable methods of research and inquiry, sets the stage for political rivalry and disputations. Having given up the notion of epistemological neutrality and objectivity in the naive sense—as if people did not count or were not involved in the production of knowledge—and having learned from the studies of sociologists of knowledge that knowledge is constructed, greater emphasis is placed on the dynamics of the scientific community.

A refusal to experiment in the postmodern sense, i.e., to try out different moves from different perspectives and with different rules, not being concerned with consistency and coherence universally but only locally, would mean that one holds onto a determined "ontology," one that through its forms of ordering in effect establishes hierarchies that in turn "legitimize oppression." As seen in Chapter 3, "It becomes obvious that there is a congruence between the domination of ontology and the torture inflicted to language games that are not those of ontological discourse" (Ibid., 53). Ontological questions are political in the sense of excluding certain possible interpretations of reality, however defined. When certain ontological commitments are made, they help establish languages in which distinctions and dichotomies are expressed and set up. When dichotomies become entrenched, they give rise to hierarchies so that when rationality is demarcated from superstition and opinion, one of the two—in most cases rationality—is preferred and elevated to a status above that of the other, and thereby becomes immune from critical engagement or translation. Though not intended to lead to this consequence, hierarchies can be easily used to oppress one language game or mode of discourse in favor of another.

If the political aspect of epistemological disputes encourages the participants to look beyond the details of their arguments not only to application opportunities but also to the very process that engenders such opportunities, then a more careful examination is required of how different discursive games can be played simultaneously as opposed to the sort of preferential treatment that one set of games, "poetry" for Rorty (1989), receives under the heading of postmodernism. (I refrain from elaborating here on Rorty's view of postmodernism and his specific version thereof because I have treated his views elsewhere: Ormiston & Sassower 1989, 115–122; Ormiston & Sassower 1991; and Ormiston &

Sassower, 1993.) What is at issue is not only the postmodern epistemological claim for new rules and games as was discussed in the previous two chapters, but a political claim for equal access to and legitimacy for a multiplicity of games.

But is reason, as defined since Plato and right into the Enlightenment, indeed pliable enough to be politicized? Is reason not reducible to the principles of logic, and therefore to a fairly rigid set of rules one either accepts or rejects completely? Perhaps it is in this respect that Lyotard admits that "There is no politics of reason, neither in the sense of a totalizing reason nor in that of the concept. And so we must do with a politics of opinion" (Lyotard & Thèbaud 1985, 82). Is opinion exclusively in the domain of normative prescriptions and reason in that of empirical descriptions? Moreover, what if reason plays only an ancillary role in scientific research and the role of opinion is significantly increased?

Raising the question about the ancillary role of reason in relation to opinion within the context of scientific knowledge still plays into the hands of a dichotomy (or binary opposition) whose usefulness may be challenged. Perhaps it is time to reconsider the privileged status and relative independence accorded to reason since Plato. This does not mean disposing of reason or restricting its usefulness; but it does mean the linkage of reason to opinion and to the pragmatics of situations so that it can fulfill its promise for communication and enlightenment, if not emancipation, without the terror of totalizing discourses. It seems to me that Martha Nussbaum shares this sentiment in her "defense" of reason not in any traditional, patriarchal sense, but rather as an important ingredient with which feminism can secure its position, arguments, and political agendas (Nussbaum 1994). One may further argue that some of those heralding the postmodern condition or any other condition of postmodernity are still reluctant to let go of the legacies of reason.

I would suggest to Lyotard, for example, that whatever he says about the study of society can be equally applied to the study of "nature" or "reality." So that when Lyotard says: "One cannot put oneself in a position of holding a discourse on the society; there are contingencies; the social web is made up of a multitude of encounters between interlocutors caught up in different pragmatics. One must judge case by case" (Ibid., 73–74), he could have said the same thing about technoscience. That is to say, the obliteration of a strict demarcation between reason and opinion is made possible by considering the question of their legitimacy as being in both cases reasonable techniques of inquiry and critique, and as such dependent on a process of self-legitimization. There is some support for this view in what Derrida says about the concept of "context." For him, the contextualization of all linguistic utterances, including those of science, is in fact a political act:

> But there is always something political "in the very project of attempting to fix the contexts of utterances." . . . What is called "objectivity," scientific for instance (in which I firmly believe, in a given situation), imposes itself only within a context which is extremely vast, old, powerfully established, stabilized or rooted in a network of conventions (for instance, those of language) and yet which still remains a context. (Derrida 1988, 136)

The contextualization process, understood by Lyotard or Derrida in political as well as linguistic terms, circumvents the (need to) appeal to some foundation (metaphysics) in order to pronounce judgments (epistemologies) about reality (ontology). It is a process that shifts the reference from an external reality or the rules of logic to conditions and circumstances internal to the context of examination. As Lyotard says:

> In other words, to respond to a case without criteria, which is reflective judgment, is itself a case in its turn, an event to which an answer, a mode of linking, will eventually have to be found. This condition may be negative, but it is the principle for all probity in politics as it is in art. I am also obliged to say: as it is in thinking. (Lyotard 1988b, 27)

As I have tried to illustrate in this chapter, "a mode of thinking" can be found: the similarities between technoscience or philosophy of science and postmodernism demonstrate the ways in which these two enterprises or groups of people can learn from each other and benefit from each other's insights. To pretend that they live in complete isolation from each other and that while one group is engaged in rational inquiry the other is lost in the abyss of irrationality is an uninformed and misleading conception that may silence potential communication and critical translation. I used here a spirit of translation (in the sense announced in Chapter 3) that may induce some to read beyond our prescribed texts and domains, no matter how difficult and annoying such an endeavor may turn out to be. I wish to carry this spirit into the next chapter, where two other discourses or language games are juxtaposed.

Chapter 5

A Feminist Engagement: Popper & Haraway

Introduction

In this chapter I return to some of the themes already developed in earlier chapters, concerning the potential of a conjunction between two discourses, both of which are critical of science, though they use different methods of argumentation and have different goals in mind. The conjunction of the concerns of Donna Haraway with "situated knowledge" and those of Karl Popper in relation to "situational logic" or "problem situation" is undertaken not for the purpose of either finding Popperian antecedents in Haraway's work or reducing their intellectual enterprises to an identical epistemological foundation. Instead, my purpose is twofold: first, to make good on my promissory note of Chapters 2 and 4 to elaborate on Haraway's ideas, and second, to provide a second demonstration of how a translation between discourses, as understood in Chapter 3, can be conducted.

I assume that both Popper and Haraway, in their own respective ways, are engaged in what has been loosely termed social epistemology, a view of the production and dissemination of knowledge through a network of institutional agencies, all of which are dynamically involved in the evaluation and consumption of a variety of forms of knowledge. I also assume that the social matrices used by either in order to contextualize their respective methodological claims are radically different; yet, despite their differences, both epistemologists deploy argumentative devices whose rhetorical force cannot remain unnoticed. As I review their views on the situatedness of knowledge, I will explore the similarities and differences that engender a tension in my attempt to translate their views to their respective colleagues.

My first hypothesis—politicizing epistemology—is that in both cases an engagement and concern with methodological issues (such as empiricism, rationality, criticism, and objectivity) express certain political commitments that respond to the dominant ideology of the scientific community (e.g., the idea that scientists can police themselves or that their research is value-neutral). That is, despite Popper's conservativism there is some radicalism in his work, even if it pales by comparison to Haraway's.

My second hypothesis—transcending situatedness—is that a Popperian contribution, however modified, can serve the feminist agenda of a successor science. That is, there is a similar concern in both Popper and Haraway to establish an appropriate scientific methodology that supersedes Baconian inductivism or the strictures of logical positivism (in some dialectical fashion) for the sake of improving the human condition (however differently they define this ideal).

My third hypothesis—situating transcendence—is that regardless of some similar epistemological concerns, the political differences between libertarian Popperians and socialist feminists (understood in both cases in Weber's sense of ideal types) cannot and will not disappear in some dialectical *Aufhebung*. That is, their different identity politics makes such a fundamental difference that to overlook it would merely resituate us in the Enlightenment world of two centuries ago, or return us to Rorty's neo-liberal position (see end of Chapter 2).

I must say, then, from the outset that every attempt to translate between a multiplicity of language games should not gloss over differences that matter, that is, differences that characterize one's position, influence one's production and consumption of knowledge, and set up the parameters within which political negotiations or demonstrations can take place.

I would like to offer some disclaimers, not merely to disarm the impatient reader as critic, but in order to shift the burden of my exploration from the rules of logic to the rules of friendly persuasion, perhaps even get us back to the metaphor of a European coffee house mentioned in the Preface. First, as already mentioned above, I have no intention to minimize any of the contributions made by Haraway—her ideas stand or fall on their own merits, and if anything, their evaluation must be contextualized within the ongoing debates of some of the feminist critiques of science and epistemology. Second, I have no interest in "rehabilitating" Popper, for such an attempt, as we have seen in dictatorships around the globe, leads to despicable hero worship or is self-serving in a patently absurd way. Third, in bringing together some ideas of these two thinkers I do not envision a "battle" from which only one victor may emerge. This encounter is a conversation whose participants may respectfully disagree with each other, while learning to make use of whatever insights are put forth (epistemological and rhetorical alike). Fourth, I hope to stimulate intellectual workers to link together many more seemingly unrelated fields of inquiry, so that blurring disciplinary boundaries will become a mission undertaken and fulfilled by more members of the academy.

Politicizing Epistemology: An Odyssey

I would like to mention my own so-called intellectual odyssey because I return repeatedly to the same texts and authors to bolster my arguments and pleas. Cultural critics of science follow in one form or another the debates of the Vienna Circle, the developments of science in the twentieth century (from the theory of relativity to quantum mechanics and beyond), and the ensuing critiques of science that were both internal (methodologists from Popper to Kuhn and Feyerabend) and external to its community (sociologists of science from Merton to the Edinburgh Strong Program and feminists from Keller to Harding and Haraway, not to mention Marxists from the Frankfurt School to Aronowitz). Besides, is it not a prerequisite nowadays to be "reflexive" and explain why one is interested in this rather than that particular problem?

My personal note may explain why the present attempt to shift from the constraints of logic to those of an appeal makes sense at all. I studied with one of Popper's disciples at Boston (Agassi) and was impressed with the radicalism of his position: Popper was an anti-inductivist in an age when (probabilistic) inductivism was the rage (from the Vienna Circle to its American stronghold in the guise of Hempel and Reichenbach); he was misunderstood and mocked more often than taken seriously; and he gave his students a bad reputation (some claim that they are unable to secure research grants). The Popperian critical edge and his dismissal of the view that Truth can ever be attained fit quite easily with my study of Marx and Marxism.

This odd couple—Marx and Popper—had a similar distrusting orientation toward the intellectual establishment of their day, and they both suffered ridicule and admiration in their respective cultures. They were, in a word, outsiders working inside a fully formulated intellectual framework, appreciating all along that the binary opposition between the outside and the inside is problematic. I know that Popper criticized Marx's theory as unscientific, but that says more about Popper's obsession with the problem of demarcation (of science from non-science) than about the presumed incommensurability of the alleged scientific models proposed by Marx and Popper. Regardless of their differences (too many to enumerate here) they exemplify a similar critical and skeptical attitude toward knowledge claims, while attempting to replace one scientific theory with a "better" one. Of course each has his own criteria according to which scientific credibility ought to be granted, but they both still remain committed to nature's "reality," to the empirical verification or refutation of theories, and to some sense of intersubjectivity. In other words, neither broke completely with Modern (philosophy of) science.

By the time I was synthesizing and applying a Marxist-Popperian critique of economics and medicine, I realized that even these radical thinkers may not be radical enough. That is, their views have a tendency to calcify and become as dogmatic as those they originally criticized. I have to admit that this sort of "consensus-building" orthodoxy is probably more characteristic of the followers of Marx and Popper than of the actual written essays and books of these thinkers themselves, but because I had access to these thinkers (primarily through the milieu of their disciples), I craved something more critical and radical. The next step in my intellectual odyssey landed me in postmodern land, a place, as we have seen so far in this book, whose landscape is difficult, if not impossible, to describe precisely. I added postmodernism to my synthesis of Marx and Popper and perceived a fractured web of critiques that could begin anywhere and end with some infusion of political awareness and at times a concern for the personal. At this point, Popper became negligible and Marx retained his honorary position backstage.

Working on science and technology from a postmodern perspective was exciting and problematic. In fact, certain views of realism and certain commitments to empiricism had to be given up in order to render full play to ideas about the construction of knowledge claims and the fictional character of all we know. Having read some feminist critiques and the works of some sociologists of science, I lost all interest in whatever dedication Popper and his followers had for their version of critical rationality. And then I read Haraway's work. Though I understand that I walked into the middle of her conversation with some feminist epistemologists and some other radical socialist feminists, I found myself suffering from an intellectual vertigo. It is not that Haraway uses the term "situated knowledge" that brought back memories of Popper's terminology, but rather that she insists on retaining certain commitments that I had long given up on because they were associated in my mind with Popper and the Vienna Circle. In short, Haraway sent me back to Popper, but now, of course, a postmodern Popper, a Marxist Popper, a feminist Popper, because I cannot read him the way I originally read him.

It is in this sense, then, that I do not plead on behalf of Popper nor do I appeal exclusively to the wisdom of Haraway. They both challenge certain views that are attractive from the perspective of French postmodernists and their American friends, yet they may carry us inadvertently toward a site that lies somewhere in the neighborhood of postmodernism. This is not to say that postmodernism is bankrupt or that its demise can be celebrated; instead, adopting the notion of displacement as opposed to replacement, I think that postmodern contributions to the study of technoscience can be contextualized in a manner that accounts for the concerns voiced by Haraway in relation to the connection of science and power and the responsibility of scientists. Displacement, as mentioned earlier, allows the cohabitation of multiple discourses, while replacement assumes that one discourse is subsumed by a subsequent one and thereby invokes hierarchical methodologies.

My brief reflections relate to my first hypothesis concerning the political nature of epistemology by illustrating the continued marginalization of the critiques of science and technology. If we grant that the dominant discourse of religion and theology of over a millennium has been displaced if not replaced by the discourses of science and technology, then it follows that despite the rhetorical gestures that seem to invite criticism, the dominant discourses of the twentieth century still ensure the marginality of their critics. The marginalization of the critiques of science expresses a politicization of epistemology, one that remains implicit in Popper, but is explicit in Haraway.

To claim nowadays that one's epistemology is either objective, neutral, or disembodied is tantamount to claiming that Truth can be ascertained by divine revelation. Epistemological debates are undertaken from the periphery in the hope that the centers of power will listen and perhaps revise their views; but these debates all too often fail to articulate their political nature. Moreover, when these debates fail to engage the centers of power they turn into peripheral debates in the periphery (Goonatilake 1993). In short, in the name of a political crusade and with the argumentative vehicles of epistemology, the responsible critic all too often forgets what the target is and what changes are indeed possible.

In adopting the displacement view of the production and consumption of knowledge claims, it is possible to retain a number of marginal views alive so that their combined articulation may have a more forceful impact on the so-called mainstream view of epistemology (the received view). The forceful impact I envision will not come from a presentation of a homogenous view; on the contrary, what will be forceful, even in the sense of having shock value, is the violation of the traditional sense of noncontradiction that may accompany the heterogeneity of the presentation. It is also possible to tease out the most useful critical tools and results that every discourse puts forth, without thereby fully adopting one discourse and completely discarding another. Following this agenda may be deemed utopian, but there is nothing shameful about hope and desire, about changing the site of epistemological debates, that is, changing the world.

I know that there are interpretive obstacles to what I seem all too cavalierly to propose here: must we return to some empirical foundation for our scientific knowledge? (More on the topic of the inevitability of reality in the next chapter.) Is the technoscientific discourse privileged enough to be the focus of critical attention and not too privileged to escape its own reformulation?

Is Haraway's vision of the "successor science" not a throwback to something quite Popperian? I say Popperian and not logical positivist because of the distinctive characteristics of Popper's ideas that, as I will show below, are much closer to Haraway's interests than to the ideas of any of Popper's inductivist counterparts. Besides, over the decades it has become less clear whether or not the lines of demarcation between induction and deduction, between rationalism and irrationalism, or between positivism and conventionalism, make any sense. The confirmation of a falsification cannot help but use induction, and the data of rigorous empiricism are still formulated in conventional manner. It is time now to examine my second hypothesis concerning the potential transcendence of situatedness.

Transcending Situatedness

Haraway stands out as the most promising and problematic contemporary critic of science because she refuses to align her position with either of the two extremes known to us: the steadfast empiricists or the fashionable postmodernists. Instead, she wishes to set up ways of validating one's scientific judgments in situation-specific circumstances so that a reliance on empirical evidence will be welcomed without thereby extrapolating from these judgments to all other situations. In her words:

> So, I think my problem and "our" problem is how to have *simultaneously* an account of radical historical contingency for all knowledge claims and knowing subjects, a critical practice for recognizing our own "semiotic technologies" for making meanings, *and* a no-nonsense commitment to faithful accounts of a "real" world, one that can be partially shared and friendly to earth-wide projects of finite freedom, adequate material abundance, modest meaning in suffering, and limited happiness. Harding calls this necessary multiple desire a need for a successor science project and a postmodern insistence on irreducible difference and radical multiplicity of local knowledges. *All* components of the desire are paradoxical and dangerous, and their combination is both contradictory and necessary. Feminists don't need a doctrine of objectivity that promises transcendence, a story that loses track of its mediations just where someone might be held responsible for something, and unlimited instrumental power. We don't want a theory of innocent powers to represent the world, where language and bodies both fall into the bliss of organic symbiosis. We also don't want to theorize the world, much less act within it, in terms of Global Systems, but we do need an earth-wide network of connections, including the ability partially to translate

> knowledges among very different—and power-differentiated—communities. We need the power of modern critical theories of how meanings and bodies get made, not in order to deny meaning and bodies, but in order to live in meanings and bodies that have a chance for a future. (Haraway 1991, 187)

For Haraway, then, it seems that there is no injunction against the strategic use of all methodological devices that have been in the service of dominant scientific culture, as long as the use is temporal and contextualized, that is, as long as it yields specific results with which to approach other problems and situations. This strategic use of the methods of science—as in the case of AIDS, for example, wherein AIDS activists master the scientific-medical vocabulary for their own use and not for the sake of buying into the ideology of the National Institute for Health or other such government agencies—may suggest what the feminist "successor science" will look like:

> feminists have stakes in a successor science project that offers a more adequate, richer, better account of a world, in order to live in it well and in critical, reflexive relation to our own as well as others' practices of domination and the unequal parts of privilege and oppression that make up all positions. In traditional philosophical categories, the issue is ethics and politics perhaps more than epistemology. (Ibid.)

This description uses terms such as "more adequate" and "better account of the world" in order to depart from the language game of postmodernists who have used terms such as labyrinths to insist that there is neither a better nor a more adequate narrative of the world (Ormiston & Sassower 1989). The question here is that of progress with its Enlightenment overtones and teleological undertones. Progress is not a term used by postmodernists, for they fear to privilege one discourse over another: they are committed to escape from having to specify permanent criteria according to which choices among discursive practices will be made. This, as some of their critics have charged, has been the downfall of postmodernists, their political bankruptcy (as already mentioned earlier, Guattari 1986 and Callinicos 1990). Yet, Haraway also uses terms, such as "radical historical contingency" and "semiotic technologies," that are closely associated with postmodern as well as Marxist rhetoric.

Postmodernists, such as Lyotard, have always claimed that "scientists, technicians, and instruments are purchased not to find truth, but to augment power." (Lyotard 1984, 46) The connection between the self-legitimating scientific discourse and the capitalist mode of production within the political power structure is neither new to postmodernists nor absent from their critiques. In this respect, the critique of science heralded by postmodernists is analogous not only to Marxist critiques but also to that of Feyerabend, the methodological pluralist, who positions himself as an anarchist in order to show the problematics of choosing between incommensurable theories and scientific models.

However, some feminist critics of science, like their radical socialist counterparts (and in some cases embodied in the same intellectual workers), demand progress in their proposals. Are they, then, adopting the Enlightenment adage of modernism that science and technology will bring about progress, that progress is rationally attainable, and that it includes the perfectibility of the human mind and the improvement of human society? You see, some modes of critique turn out to be modernist through and through, because the repeal of modernist standards illustrates a certain level of complicity, a form of paying homage to what is significant and needs revision. In this sense, then, a feminist-postmodern critique tries to avoid this pitfall. Haraway, for one, sidesteps this "progressive-modernist" trap for it bespeaks of hegemony and dominance, uniformity and conformity, and last but not least, it harbors the conditions for oppression.

Haraway does not mind using the traditional terms of science and technology, but she insists that they be defined differently or at least defined by feminists, however differently they view themselves and their theoretical apparatus. For the power to define is all important in political and epistemological debates alike. What other words are used, then? Haraway talks about politics and ethics in ways that need not be collapsed into an epistemological debate about truth, for her introduction of a normative component at this juncture of her discussion complicates the situation so that a simple equation of progress and truth and ethics cannot be formulated.

Part of the complexity of the triad of epistemology, politics, and ethics has to do with, as one would expect from feminist critics of science, the power relations that pervade the scientific community and that define its role within society. So, instead of holding onto classical definitions of epistemology and truth, of science and objectivity, Haraway says:

> Feminist objectivity is about limited location and situated knowledge, not about transcendence and splitting of subject and object. In this way we might become answerable for what we learn how to see. (Haraway 1991, 190)

Having inserted the researcher into the equation that eventually yields answers to questions we ourselves raise, Haraway goes on to say that "feminist objectivity means quite simply *situated knowledges*." (Ibid., 188) What is this situated knowledge (whether in the singular or the plural)?

Though standpoint theory is commonly associated with Harding and has its own elements, Haraway seems to view her version of situated knowledges in light of it so that there are sufficient common elements to warrant parallels if not equivalence. According to Haraway: "the feminist standpoint theorists' goal of an epistemology and politics of engaged, accountable positioning remains eminently potent. The goal is better accounts of the world, that is, 'science.'" (Ibid., 196)

Haraway's voice is added here to other feminist voices that demand to be heard, that insist on having their vision receive the legitimacy accorded previously only to male-dominated scientific inquiries, and that appeal to as wide an audience as is possible. The "units" of analysis are not limited to inanimate objects and entities, but include humans in a multiplicity of roles and their ideologies, belonging at different times to a variety of groups and associations.

For Haraway, as for Toulmin, the shift from the scientist as spectator to the scientist as participant is linked not only to epistemological correctness but to an ethical awareness, political conviction, and social responsibility (Ibid., 198). As an important consideration that is strangely but all too often ignored, it may be useful to recall here that the feminist concern with responsibility is not completely absent from the work of Lyotard (as seen at the end of Chapter 3), who says, in relation to the multiplicity of language games, that "it makes the 'players' assume responsibility not only for the statement they propose, but also for the rules to which they submit those statements in order to render them acceptable." (Lyotard 1984, 62)

Situating Transcendence

How does the scientist as critic go about formulating knowledge claims? No longer is it possible to assume with Bacon that observations (more precisely, observation reports or statements) can be collected into generalizable knowledge. Nor is it tenable to believe that science is nothing but the approximation of some universal Truth. This much even Popper admits in his tenacious attacks on inductivism, while proposing an alternative scientific method of conjectures and refutations. According to Popper:

> The difference between the amoeba and Einstein is that, although both make use of the method of trial and error or elimination, the amoeba dislikes erring while Einstein is intrigued by it: he consciously searches for his errors in the hope of learning by their discovery and elimination. The method of science is the critical method. (Popper 1979, 70).

What is the so-called critical method? How is this method applicable to all the problems studied by scientists? Popper's reply is the following:

> By a situational analysis I mean a certain kind of tentative or conjectural explanation of some human action which appeals to the situation in which the agent finds himself. It may be a historical explanation: we may perhaps wish to explain how and why a certain structure of ideas was created. Admittedly, no creative action can ever be fully explained. Nevertheless, we can try, conjecturally, to give an idealized reconstruction of the problem situation in which the agent found himself, and to that extent make the action "understandable" (or "rationally understandable"), that is to say, *adequate to his situation as he saw it*. This method of situational analysis may be described as an application of the *rationality principle*. It would be a task for situational analysis to distinguish between the situation as the agent saw it, and the situation as it was (both, of course, conjectured). Thus the historian of science not only tries to explain by situational analysis the theory proposed by a scientist as adequate, but he may even try to explain the scientist's failure. (Ibid., 179)

In this sense, then, Popper's scientist—though it remains the male scientist throughout his works—is cast into the same role as envisioned by Haraway (for whom scientists may be either male or female, and preferably cyborgs). The scientific conjecture is contingent: it is historically conditioned and as such cannot make any universal or transcendent claims. Once again, this notion is similar to the concern Haraway has over the totalizing nature of any knowledge claim and the assemblage of such claims into a successor science. As she says: "The feminist dream of a common language," may it be the language of science or of aesthetics, "like all dreams for a perfectly true language, of perfectly faithful naming of experience," as was the case with the introduction and application of Latin and of mathematics, "is a totalizing and imperialistic one." (Haraway 1991, 173) If there is a move away from totalizing explanations toward historically circumscribed ones, what will be lost? What distinctive "scientific" character of knowledge claims will be abrogated? Moreover, how is this move different from that of the postmodernists?

Popper is worried, just as Haraway is, that the participant scientist will devalue the thrust of an explanation because it is "simply" historical in the sense of being a subjective claim that can be trivialized and dismissed offhand. As Popper says: "Thus what he has to do *qua* historian is not to re-enact past experiences, but to marshall objective arguments for and against his conjectural situational analysis." (Popper 1979, 188) To ensure some sense of objectivity, there must be a grounding that appeals *a priori* to some social common sense that goes beyond the enclosed parameters of the investigator. For, as both Popper and Haraway would agree, to have any political credibility and applicability, explanations must be translatable in my terms.

There is also the problem of shifting attention from the scientists to the science critic and eventually to the historian of science. It makes perfect sense to insist that the scientist also be a critic and a historian of science; but unfortunately it is difficult to assume this situation, especially given the educational procedures undertaken by future scientists (more on this in Chapter 7). So, when Popper shifts his discussion from the scientist to the historian, it is unclear if they are meant to be the same person or not. To quote Popper:

> The historian's task is, therefore, so to reconstruct the problem situation as it appeared to the agent, that the actions of the agent become *adequate* to the situation. This is very similar to Collingwood's method, but it eliminates from the theory of understanding and from the historical method precisely the subjective or second-world element which for Collingwood and most other theorists of understanding (hermeneuticists) is its salient point.
>
> Our conjectural reconstruction of the situation may be a real historical discovery. It may explain any aspect of history so far unexplained; and it may be corroborated by new evidence, for example by the fact that it may improve our understanding of some document, perhaps by drawing our attention to some previously overlooked or unexplained allusions. (Ibid., 189)

Popper's investigator, who is historically informed, must reconstruct the situation by adding cultural variables regardless of the "logic" of the situation. Because I find the demarcation between the sciences both problematic and contestable, it may be helpful to consider that all sciences are social sciences by definition, sciences whose social context is acknowledged through the hallmark of social epistemology with all its variants. From this perspective, Popper's concerns with the application of situational logic in the social sciences can be applied to all other fields of alleged scientific research. Yet, even for Popper, the reconstruction need not be reduced to the articulation of the psychological components of the situation. Instead, the concern with individual behavior and reaction, decision-making and evaluation, can be accounted for logically. The use made of logic in this context seems more pliable than in other contexts. According to Popper:

> The method of applying a situational logic to the social sciences is not based on any psychological assumption concerning the rationality (or otherwise) of "human nature." On the contrary: when we speak of "rational behaviour" or of "irrational behaviour" then we mean behaviour which is, or which is not, in accordance with the logic of that situation. In fact, the psychological analysis of an action in terms of its (rational or irrational) motives presupposes—as has been pointed out by Max Weber—that we have previously developed some standard of what is to be considered as rational in the situation in question. (Popper 1966, 97) (See also Popper 1957, Sections 31 & 32 on the application of situational analysis/logic to history).

Even Popper the methodologist is a sociologist of science. In his concession to the significance of social factors that establish epistemological claims and theories, Popper seems to favor the notion of situation-oriented modelling of scientific data and explanation. In this regard, his view can be connected with the view of some social constructivists, such as David Bloor:

> All knowledge, the sociologist could say, is conjectural and theoretical. Nothing is absolute and final. Therefore all knowledge is relative to the local situation of the thinkers who produce it: the ideas and conjectures that they are capable of producing; the problems that bother them; the interplay of assumption and criticism in their milieu; their purposes and aims; the experiences they have and the standards and meaning they apply. (Bloor 1991, 159)

This much has been agreed upon by most critics of science in the 1990s, regardless of whether they consider themselves radical or not: the finality and universality expected of the scientific discourse has been shattered once and for all; and in this respect, the successor science cannot reclaim this posture; instead it has to construct a different mode of operation, one whose appeal will bring together a variety of critiques.

Rationality as Situated Logic

My third hypothesis concerns the situatedness of transcendence, that is, the differentiation between the Popperian and feminist proposals for contextualizing epistemology while still maintaining the potential for intersubjectivity or for the credibility of judgments. Now, if Haraway shifts the focus of science from epistemology to ethics and politics, is there no danger that the best one can do in a given situation is provide a personal account of the circumstances as seen by an individual? How does one formulate a political agenda that is not limited to personal pleas? Will it suffice to remain in contact with one's political community? Moreover, will the scientific appeal become (despite Popper's conditions) psychologically grounded so that one's own ego will turn out to predetermine what explanation is rendered in a specific situation?

Haraway does not want this result, nor is it a result that would buttress the claims of a feminist successor science against the criticisms of non- or anti-feminists. Haraway's appeal recognizes the paradox of being at once a successor and an alternative, that is, a recognition that a multiplication of technoscientific discourses could still marginalize the import of a feminist-oriented science that is supposed to replace, displace, or at least improve upon all previous attempts. Must Haraway (and by extension anyone else interested in the development of a successor science) revert to the use of some form of rationality to state her case so as to appeal not only to a fragmented set of feminist critiques but to all other marginalized critiques of science and technology?

There is no need to rehearse the critical literature concerning rationality as the logocentric basis of modern science. Yet, it is crucial to recall that Popper, unlike some feminist critics, remains indebted to some principle of rationality. And in this indebtedness one finds the unrelenting and dialectical tension between Popper and Haraway. For Popper, rationality is the way to intersubjectivity, perhaps because it is too much to expect objectivity. Would Haraway be satisfied with this epistemological grounding? Anything that would be a simple or full throwback to the masculine Age of Reason would be rejected out of hand, even when some of the issues of that Age remain important today. Still, Haraway readily concedes that "the science question in feminism is about objectivity as positioned rationality." (Haraway 1991, 196) That is, as long as rationality is qualified and redefined by feminist critics, it remains a useful tool in the hands of some Popperians and some feminists alike. But is it the same rationality? Even if not, is there a way to redefine rationality in less hegemonic or oppressive terms so as to translate (between) Popper and Haraway, or perhaps acknowledge their concern with social and political connections from which decisions can be made?

Perhaps even a minor concession to some level of rational intersubjectivity as a political move toward a forum that permits critical evaluation and debate may illustrate the dialectical relation between Popper and Haraway. This concession recalls Nussbaum's concern (Chapter 3) as well as my first hypothesis about assembling a multiplicity of politically marginalized critiques of science for a more effective political rhetorical force for the sake of social change. According to Agassi:

> Viewing science chiefly as purposive or as goal-directed activity is much more in accord with the sociological approach favored by Popper, namely the rationality principle, or situational logic, or the principle of rational reconstruction. The word "rationality" in the sense of puposiveness should not be confused with "rationality" in the sense of enlightenment. (Agassi 1975, 328).

Though Agassi's interest lies with "the sense of enlightenment" that he equates with the notion of science and scientific explanation, one can find here a process of relativizing rationality by contextualizing its meaning for particular purposes. While the notion of purposiveness would appeal to Haraway more than that of enlightenment, it is reasonable to assume that Popper is more interested in scientific enlightenment and thus finds the notion of purposiveness to be a limiting case of the principle of rationality.

The notion of situational logic or the logic of situations may turn out to be the methodological or argumentative device whereby an epistemology of sorts will contain ethical and political considerations as a matter of course and not as *post hoc* additions. As such, this methodological turn may foreground the differences between critics like Popper and Haraway as opposed to whitewash them. Just because some Popperians and some feminists who agree with Haraway adopt a similar vocabulary (of situational logic and epistemology) does not mean that their use of the terms and methods will yield similar results. Why?

Ian Jarvie considers the Popperian situational logic to have been one of his most important contributions to science, even though this is the least documented of his views (found primarily in lectures, says Jarvie). Though the parameters of explanation turn on two additional principles (to that of rationality), the principle of causality (despite David Hume's problematization) and the principle of means/ends analysis, Jarvie still finds situational logic the most appropriate means with which to provide an objective analysis or explanation of a social outcome. According to Jarvie:

> It is assumed that the situation, if objectively appraised, should favour certain means which are more effective than others and that the measure of rationality consists in the success in approaching such an "objective" appraisal. The logic of the situation, then, is an empirical description of the procedure of explanation which goes on in the social sciences; it is also a normative prescription for reform of what does not fit the description, particularly holistic and psychologistic social science; it is also, finally, a logical analysis of what underlies plausible social science explanations. The logic of the situation is a special case of the deductive

> analysis of causal explanation in general and illustrates the unity of method in the sciences. (Jarvie 1972, 4–5)

Regardless of how upbeat Jarvie sounds and despite his nostalgia for a "unity of method in the sciences" (reminiscent of the ideals of the Vienna Circle), he is willing to admit that there are some problems associated with the application of this method. In his words:

> the fundamental model for explanation in the social sciences is the logic of the situation. The logic of the situation has deeply embedded in it the rationality principle, the methodological rule that we attribute rationality—goal-directedness—to human actions in need of explanation, unless we have good grounds for not doing so. What we find problematic, and what we find rational, are tied in essential ways to our own society and its outlook. (Ibid., 66)

Here, then, is the locus of the tension between the explanatory devices proposed by Popper and Haraway, a tension that arises in some of their own words and perhaps more noticeably in the words of some of their critical commentators: in order to contextualize the logic of a situation an appeal must be made to culture. But, as Fredric Jameson aptly reminds us, "in postmodern culture, 'culture' has become a product in its own right" (Jameson 1991, x). Once such an appeal is made, it is no longer possible for just anyone to enter a situation as though he or she were in a thought experiment that provides the parameters of decision-making and a model of explanation. The backdrop or stage is itself reconfigured as the performance takes place. As Haraway would insist, the personal background of oppression and discrimination, of values and ideals with which one encounters the world and is encountered by others, constructs one's identity in problematic ways, but ways whose difference remains undeniable. The insertion of a (Popperian) disembodied logician, one who admittedly is trying to avoid the pitfalls of psychologism, is so unreasonable and untenable that it turns out to be (epistemologically) useless.

It is not only that the Popperian situational logic turns out to be too logical and not personal enough, but it presumes unreflectively that the definitions of rationality pertinent to particular situations are uniform with regard to some cultural background knowledge. This (political) posture belies one's confidence that naming and defining can be done objectively, that is, undertaken from an Archimedean perspective and not from the messy perspectives of people who are engaged in the situations that they need so desperately to explain. Without those explanations their voices would not be heard and their visions would be obscured.

Would it be possible for someone from this perspective to imagine situations and contexts different from those they have been working with? I attempt to give a partial answer to this question in my explorations in Chapter 7.

The Politics of Knowledge: Strategic Alliances

When surveying the respective literatures of some of the Popperians and some of the feminists like Haraway, it is evident that the two groups do not read each other's works, or if they do they persist in not acknowledging their respective contributions. Now one can claim that this situation is yet another instance of academic professionalization wherein one is hardly capable of keeping up with the literature of/in one's own field of specialty. But this explanation is not good enough, for there are feminist critics of science who should know of Popper, and there are Popperian sociologists of science who should know at least of Keller, Harding, and Haraway. So, to keep abreast of one's own specialty would require a much broader intellectual reach of these two groups, one that would refer to works emanating from either camp. Why is that still not the case? Why am I attempting to translate, learn and relearn, and appeal to more than one audience at a time?

Here is where I return to the theme of this book, a theme that politicizes the process of situating knowledges. The politics I have in mind is threefold. First, as has been illustrated from the sources I quoted above, to think of knowledge claims that are made in relation to a particular situation is to invoke a set of parameters that are neither objective in some traditional sense nor value-neutral in any sense of the term. So, whatever argumentative claims are made in regard to a situation, whatever logical moves are deployed to provide an explanation, these moves and explanations are culturally contextualized (unless they are meant to be transcendent moves). The selection of the cultural criteria according to which judgments are made is politically informed and should be consciously exposed, the choice of variables that are foregrounded or marginalized is politically motivated, and the kind of explanation that results from the analysis is politically potent.

Second, there is a political standpoint from which critics of science speak and to which they are committed either implicitly or explicitly. Popper's political views are considered conservative and libertarian in nature because of his preference for a *laissez-faire* economic world and the piecemeal engineering of the social sphere over revolutionary changes. Yet, I find his insistence on the method of conjectures and refutation a liberating move from the oppression of inductivist collection of data and the generalizations that are derived by the leaders of the scientific community. As far as the Popperian principle is concerned, anyone can (in principle) make any conjecture and the entire scientific community is responsible to find falsifying instances for that conjecture. Of course this description is quite simplistic and naive for it fails to specify sociologically whose conjectures are, in fact, tested. But the possibility remains.

Haraway, by contrast to anything Popperian, is clear about her feminist and socialist agenda, and makes no excuses for setting a political agenda that attempts to revolutionize not only the scientific community as an institution, but the entire scientific edifice as an ideology. That Haraway's politics differs from Popper's is unquestionable; but what may still be of interest is the possibility that the Popperian method may allow a sense of individual empowerment and social change whose political force may turn out to be instrumental to a radical feminist agenda. In what sense? To begin with, Popper relinquished the quest for Truth as it was codified in the beginning of the twentieth century. Claiming that all we can know for sure is what is false remains an important epistemological statement of a critical skeptic whose aim is to learn more about the world, not less. This epistemological position is political through and through, for it allows the meekest of voices to falsify any fortified theory; theories are hypotheses, and hypotheses are conjectures, and even when they withstand criticism their truth claims remain putative at best. Security and power, stability and status, are no longer taken for granted in this epistemological and political worldview, but are in fact diminished, if not relinquished forever.

Third, there is the politics of the academy. Intellectuals feel the need to express their political allegiances, usually traceable to their graduate days, their mentors, peers, and ideological colleagues, their universities, and the communities to which they choose to belong (because of their activism or lack thereof). When intellectuals get jobs in the academy, they are pressured to swear their allegiances to their departments and the professional organizations to which they must belong to ensure tenure and promotion. But at some point this entire political mapping becomes transparent even to its participants, and we have to begin asking questions about our political commitments. As engaged intellectuals, we should fight about destiny and the future, about our utopian vision and the voices that will carry it along, but not about ancestry and originality or about prestige and status.

Perhaps with this plea in mind, revisiting the concern of Popper and Haraway with the limits of and the need for objectivity and intersubjectivity makes sense. Notice that I refrain from reconciling their epistemological differences and their political convictions, because such a reconciliation will be a throwback to liberalism of the most reductionist kind. Instead, in the next chapter I continue my exploration into the material conditions under which all critiques may have to be produced. The next chapter does not rehearse neoclassical economics or some version of neo-Marxism. Instead, it is a recognition of the importance of paying attention to the conditions under which a postmodern orientation in the age of technoscience may take shape.

Chapter 6

The Material Conditions of Postmodern Technoscience

Introduction

In the first part of this chapter, I account for the inevitability of reality or the quest for "real" foundations as it is exemplified in parts of the previous three chapters. It is of considerable concern to have some reference point against which to measure one's contentions, whether they are considered scientific or political in nature. What will become evident in this chapter is not necessarily the inevitability of reality in some ontological or metaphysical sense, but rather a more modest appreciation of the material conditions under which the practices of technoscience are undertaken. If understood appropriately, this shift in the discussion will explain some of the issues already raised in Chapter 1 in relation to the SSC project. If one wishes, the present chapter is thereby a neo- or post-Marxian attempt to concretize the philosophical concerns—understood in postmodern terms—underlined in the previous chapters.

I argue here that it is a grave oversight to conflate knowledge claims about reality or their truth status (and validity) with the very existence of that same reality. Of course, this reality is not limited to inanimate objects, but includes the social reality that is culturally constructed. I argue here that no matter how humanly limited all knowledge-claims about reality are and how their truth status may remain forever putative at best, it is unreasonable (as opposed to irrational) to deny the existence of the reality with which humans routinely engage and of which they are an integral part. It is in this sense, then, that the inevitability of reality is neither a Faustian nor a Sisyphusian curse, but instead a bedrock of psychological stability and social communication. The inevitability of reality is an invitation to explore, interpret, and imagine whatever there is and how it should be shared epistemologically, politically, and financially. Translating the epistemological insights of positivists and social constructivists alike, without thereby avoiding critical debates and fundamental disagreements, would enable society to reconceive its fortunes and fight for its dreams and aspirations.

The second part of this chapter moves from a general discussion of the need to reconceptualize theories that traditionally come under the label of political economy, to a more specific analysis of corporate structure and culture in the information age. The argument is not limited to comprehensive proposals concerning the need to revise old-fashioned and outdated neoclassical models and principles. Instead, it tackles the difficult restructuring of corporations in a changing cultural climate that defies the classical categories of capitalism and socialism. In many ways this chapter outlines an agenda for future scholars and practitioners rather than providing either a description of the present or a detailed blueprint with an attached budget. It is, therefore, an invitation for critical negotiation and not an injunction; an invitation all too seldom extended to so-called outsiders.

The Inevitability of Reality

Assume with Haraway that: "Nature cannot pre-exist its construction, its articulation in heterogeneous social encounters, where all of the actors are not humans and all of the humans are not 'us,' however defined . . . . Nature is a topic of public discourse on which much turns, even the earth." (Haraway 1992, 67) But of what does this so-called public discourse consist? Who defines the parameters, the framework, the linguistic apparatus? Besides, does it make any sense (common sense, if you wish) to conceive of the idea that nature, however defined, depends on human construction for its existence? We may not need to quote God's admonition of Job: "Where was thou when I laid the foundations of the earth?" (Job 38:4ff), nor need we be religiously inclined to appreciate the preposterous pretense of generations of intellectuals who have maintained that nature, reality, or the universe needs human language or legitimation for its existence. If, instead of existence, these same people talked about the linguistic articulation of said existence, there would be no outrage and no problem. But this, of course, is a trivial point: who would deny that the articulation of an idea or phenomenon depends on language? The claim under consideration here must therefore be quite different indeed.

Haraway is admirably honest when she concedes the less trivial point concerning the problem of and need for human articulation of nature's existence (already quoted in the previous chapter): "So, I think my problem and 'our' problem is how to have *simultaneously* an account of radical historical contingency for all knowledge claims and knowing subjects . . . *and* a non-nonsense commitment to faithful accounts of a 'real' world." (Haraway 1991, 187) The concession, as I call it, is in fact an admission that the epistemological project of some feminists, radical-socialist or others, is such that it requires an adherence to, or recognition of, some form of ontology. In short, any epistemological project—feminist or old-fashioned positivist—must be concerned, at some level of articulation, with its presuppositions concerning ontology.

It is not that Haraway "discovered" the problems associated with knowing reality, nor is she the only contemporary epistemologist to pay attention to them. On the contrary, what makes Haraway's formulation so interesting in the late twentieth century is her "return" to themes articulated over the centuries by philosophers of science and scientists alike. Haraway resists the fashionable flight from ontological commitment that is arguably displayed by some postmodernists and some standpoint epistemologists. For example, Pierre Bourdieu equates realism with science and disdains the "nihilistic attack on science, like certain so-called 'postmodern' analyses, which do no more than add the flavor of the month dressed with soupçon of 'French radical chic' to the age-old irrationalist rejection of science" (Bourdieu 1988, xii–xiii).

In Haraway's admission one sees beyond the self-reflexive moves of some postmodern critics and their disciples into a historically rich past of science and the philosophy of science. As such, some of the comments quoted above echo the concerns of Einstein and Russell, earlier in this century, about the relationship between "naive realism, i.e., the doctrine that things are what they seem," as Russell says, and physics, the science of the "effects" of phenomena on their observers. (Agassi 1975, 119)

It is characteristic of modern philosophy, since the days of Bacon and Descartes, to be concerned with epistemological questions as they relate to ontological convictions. They all agree on the existence of reality, the world surrounding them, or nature. But what are the alleged essential characteristics or components of these categories or entities? The standard demarcation between the empiricists and the rationalists rests on the distinction between knowing the truth about reality either through the senses or a priori (through the logical operations of the mind alone). The former method or procedure of knowledge acquisition is also known as sensationalism, while the latter is known as apriorism. These two epistemological methods are then also linked to the inductivism of data collection on the one hand, and to the deductions of logical rules on the other. Likewise, these two methods of inquiry are also distinguished in terms of realism (admittedly naive) and idealism. But if Russell is correct about naive realism in its encounter with twentieth-century physics, then naive realism slips into some sort of idealism quite readily: it denies the direct correspondence between the reality of an object and the observation-report of that object.

Likewise, as any student of logic knows, in a valid deductive argument, if the premises are true the conclusion must necessarily be true. But it is quite problematic to assess the truth status of general premises because their formulation depends on the observation reports and empirical data that make up an inductive argument. In short, the premise of a deductive argument is often the conclusion of an inductive argument. Where do we go to from here?

The Ambiguity of Reality

The closer scientists get to the reality they study, whether that of astrophysics or that of subatomic substances, the farther away they seem to drift from simple answers to ontological and metaphysical questions. In their eyes, their apologists hasten to claim, they are not drifting at all, but merely gesturing toward the inevitable truth: all their epistemological claims are nothing more nor less than approximations of the reality or the truth about the reality they study. With this caveat they all seem to live comfortably, so that the charges of less informed philosophers of science and sociologists of knowledge are deemed negligible, if not completely misguided. Scientists do the best they can, given the state of the technology at their disposal, and intellectual worries and speculations can safely remain with astrologers, poets, and philosophers.

Whether one extrapolates the attitude of scientists from their practices, as Latour does with the laboratory (Latour & Woolgar 1986); from their published scientific accounts, as Agassi does with Faraday (Agassi 1971); or from their philosophical musings (e.g., Einstein (1954) on Peace and God), one thing becomes clear, as E. A. Burtt argues at the beginning of the century: all scientific views, theories, and models are based on and imply certain metaphysical views and theories (Burtt 1982). The epistemological link with ontology ends up metaphysical through and through. Now, what does it mean to introduce metaphysics at this juncture? Following Hume's lead on the problems associated with the use of induction as a method for knowledge acquisition, Popper, for one, "showed that, in spite of confirmationist hopes, no amount of observational evidence could ever increase the probability, let alone guarantee the truth, of a given hypothesis." (Gatens-Robinson 1993, 538)

Once metaphysics is entered as a variable into the epistemological equation—whether an individual believes in the efficacy of logical statements as neutral devices with which to attach truth-values to empirical reports or believes in the inherent theory-ladenness of all observations and all the languages with which these observations are formulated—there is a different sense of the truth status of knowledge claims about reality (see also Chapter 3). Is reality indeed uncovered and discovered, or is it constructed by humans for particular purposes under particular circumstances and with specific prejudices or presuppositions?

Though dealing with the realities of the existential movement in France and Nazi atrocities during World War II in concentration camps, I think it useful to adopt Simone de Beauvoir's terminology in trying to answer the question just posed. For instead of having to choose between the two extreme positions contained in the question, either the fact of reality in its most universalizable and durable sense or the pliability of reality in its most subjectified and constructed sense, perhaps there is a way of finding a middle position or superseding the dichotomy altogether. In de Beauvoir's terms, instead of giving in to the paralyzing posture of absurdity (since, in our case, either option is too extreme and therefore untenable), she advocates an engagement with ambiguous situations. She says: "The notion of ambiguity must not be confused with that of absurdity . . . to say that it is ambiguous is to assert that its meaning is never fixed, that it must be constantly won." (de Beauvoir 1991, 129)

In order to ensure a dialectical tension and not commit to a middle ground, compromise, or non-contradiction, it is useful to remain critical or skeptical toward claims about reality, while still making these claims with passion and conviction. This is what I tried to explore in the previous chapter when relating Popper's notion of situated logic and Haraway's critical evaluation of the notions of situated knowledges. Whether labeled critical realism (e.g., Murphy 1990) or the epistemological method of the feminist successor science, it seems that de Beauvoir, in her own way, recognizes the need to embrace ambiguity as an intellectual and practical challenge. There is no sense in adopting the nihilism of absurdity (the irrationalism of the Romantics in the previous century and the supposed stance of the postmodernists of this century, as explored in Chapter 4) or the dogmatism of empiricism (the religious zealotry of past centuries or the positivist convictions—empirical, logical, and statistical—of this century). Instead, all epistemological pronouncements and their practical settings are bound to be simultaneously transparent and opaque, always ambiguous.

I will return later to de Beauvoir, Popper, and Haraway. But before I proceed, I wish to recapture the metaphysical thrust of any discussion about reality by quoting, as is the habit of philosophers, some of the fragments attributed to the preSocratic thinkers. This, in turn, would not only provide a historical setting for the concerns of modern and contemporary philosophy of science, but would also provide a reference point for the formulation of the metaphysical convictions I believe will guide most of the intellectual community in the twenty-first century.

PreSocratic Reality

To appreciate the epistemological debate between the empiricists and rationalists of the past three hundred years and the debates between contemporary empiricists and constructivists, it may be useful to return to the debates of the preSocratics. The return or recollection of preSocratic fragments is unlike the Popperian attempt to recast them within his framework so that they appear now to have been "Popperians" all along, perhaps proto-Popperians (Popper 1963, Ch. 5). Instead, the issue is still epistemology and its linkage to ontology in the broadest sense, not limited to the modern conceptualization of science as the only legitimate venue for epistemology (as opposed to mythology, theology, visual arts, or music). It seems to me that at least two major metaphysical concepts or principles are at stake, perhaps two metaphysical "desires": (1) a quest for *order*, an existential angst, a psychological drive, and a personal journey toward the safety of the assurance that this metaphysical goal is within human capacity or reach; and (2) a quest for social or communal recognition, and even legitimacy, of one's *rational* explanation of any phenomenon. Now, of course, (1) and (2) are related: for the order of the universe is perceived to be rationally explicable, and all rational explanations are universalizable so that they provide an order beyond the idiosyncrasies of any individual manifestations.

The preSocratic literature is fraught with a skeptical attitude toward the attainment of truth through the senses, despite the realization that without the senses it would be impossible to talk of any truth about nature. What should be done? Should the preSocratics adopt the language of metaphors, allegories, and myths, or should they adhere to the rigors of logic? As Kirk, Raven, and Schofield argue: "The transition from myths to philosophy, from *muthos* to *logos* as it is sometimes put, is far more radical than that involved in a simple process of depersonifying or demythologizing, understood whether as a rejection of allegory or as a kind of decoding." (Kirk et. al. 1983, 73)

It seems that the ancient radicalization of myth and philosophy includes a range of discursive techniques that would obliterate and undermine some of our cherished distinctions between fact and fiction, science and pseudo science, reality and the imagination. And this, of course, comes close to the portrayal of the postmodern orientation in this book.

Anaximander of Miletus constructs a theory of the origins of humans that comes closer to what contemporary scientists call evolution. As the fragments say:

> **133** Anaximander said that the first living creatures were born in moisture, enclosed in thorny barks; and that as their age increased they

came forth on to the drier part and, when the bark had broken off, they lived a different kind of life for a short time.

134 Further he says that in the beginning man was born from creatures of a different kind; because other creatures are soon self-supporting, but man alone needs prolonged nursing. For this reason he would not have survived if this had been his original form.

135 Anaximander of Miletus conceived that there rose from heated water and earth either fish or creatures very like fish; in these man grew, in the form of embryos retained within until puberty; then at last the fish-like creatures burst and men and women who were already able to nourish themselves stepped forth. (Ibid., 141)

It seems reasonable to concur with the editors of the preSocratic fragments that "Anaximander's is the first attempt to explain the origin of man, as well as of the world, rationally . . . Incomplete as our sources are, they show that his account of nature, though among the earliest, was one of the broadest in scope and most imaginative of all." (Ibid., 142) It is the conjunction of rationality, order, and imagination that transcends the confines of historical reconstruction and throws light on the potentialities of twenty-first-century thinkers. Indeed, the punctuated evolutionary theory of Stephen Jay Gould, for instance, is no more nor less imaginative or scientific than Anaximander's.

Xenophanes has the following to say about the limitations of human knowledge:

186 No man knows, or ever will know, the truth about the gods and about everything I speak of; for even if one chanced to say the complete truth, yet oneself knows it not; but seeming is wrought over all things [*or* fancy is wrought in the case of all men].

187 Let these things be opined as resembling the truth . . .

188 Yet the gods have not revealed all things to men from the beginning; but by seeking men find out better in time. (Ibid., 179)

Is it indeed a question of time till we ascertain the truth, the whole truth, and nothing but the truth? Or is it rather an untenable proposition to even conceive of coming across the truth? And, as Xenophanes so well says, even if we did come across the whole truth, how would we know? What signs would assure us of our criteria? These questions pull us back from ontology and metaphysics to epistemology.

Most often it is Heraclitus of Ephesus who has been perceived to be the champion of the view that the nature of knowledge is enigmatic. His view has been quoted in terms of the impossibility of crossing the same river twice, and the view that all things in nature are in flux. (Ibid., 195)

Whether Heraclitus was professing the common knowledge of daily experience or profound epistemological principles remains an open question. Yet, what remains indisputable is the general intellectual (for lack of a better term) atmosphere of the time, one that is sensitive to the inherent problems of validating and legitimating any and all knowledge claims. This atmosphere, as we shall see later, is premodern in Latour's sense.

Anaxagoras of Clazomenae is more specific than Heraclitus about the inevitable deception of the human senses. In his words: "**509** From the weakness of our senses we cannot judge the truth. **510** Appearances are a glimpse of the obscure." (Ibid., 383) His contemporary, Melissus of Samos, echoes the rejection of the reliance on the senses when he says:

> We said that there were many things that were eternal and had forms and strength of their own, and yet we fancy that they all suffer alteration, and that they change from what we see each time. It is clear, then, that we did not see aright after all, nor are we right in believing that all these things are many. (Ibid., 399)

Democritus of Abdera, known as the canonical figure to have argued for atomism quite a bit later than Heraclitus and Melissus, is forceful in his skeptical attitude toward the acceptance of reports based on the senses. As he says:

> Now that in reality we do not grasp what each thing is or is not in character, has been made clear in many ways . . . A man must know by this yardstick that he is separated from reality . . . This argument too shows that in reality we know nothing about anything; but for each of us there is a reshaping—belief . . . Yet it will be clear that to know in reality what each thing is in character is baffling. (Ibid., 410)

Elsewhere he says: "Wretched mind, do you take your assurances from us and then overthrow us [*sc.* the senses]? Our overthrow is your downfall." (Ibid., 412)

One need not be a Whiteheadian and paraphrase his dictum to say that the history of the philosophy of science is nothing but footnotes to the preSocratics in order to appreciate the great insights and concerns voiced some twenty-five hundred years ago. The overthrowing of the mind is the downfall of its great promise and appeal; it also illustrates to what extent it is true that the knowledge of nature acquired by humans remains baffling. Yet, a skeptical attitude toward epistemology need not lead one to assume that there is no reality whatsoever, or that the only way to understand anything is through rational discourse devoid of the reports of the senses. On the contrary, Harding is correct when she reminds us of the importance of science as a way to be literate about one's reality and culture: "we live in a scientific culture; to be scientifically illiterate is simply to be illiterate—a condition of far too many women and men already." (Harding 1991, 55)

Perhaps the social constructivists and the logical positivists can remind each other of the atomism of Democritus and the great effort it took twentieth-century scientists to rid themselves of centuries of belief in "ether" as the medium through which rays of light moved and in which atomic interactions took place. This is not to suggest that postmodernists, for example, are no different from preSocratics; instead, my brief recapitulation suggests that the problems associated with the acquisition of knowledge about reality are perennial. Moreover, the displacement of discourses would have ensured that preSocratic fragments would be part of the vocabulary used and appealed to by some postmodernists and feminists who work in the age of technoscience.

Reality, Politics, and Ethics

If all knowledge claims are inherently incomplete or at best conjectural, if no truth ("Truth") can ever be fully ascertained, and if the incomplete conjectural truths we establish are context-bound, as Popper, de Beauvoir, Lyotard, and Haraway all agree, what are we to do? Perhaps we must concede to the (epistemologically) frightening proposition of Michel Serres that:

> All things are emptied of their reality through rivalry. Every science is void of its truth through rivalry. You who fight for your truth possess only the truth of the contest. You who fight for knowledge possess only the knowledge of the battle. Soon, there will be only one science, the science of battles. The science of all sciences will only be an immense strategy, the space of knowledges lies in the hands of society. (Serres 1989, 48)

Serres alludes to an even more problematic situation, one that transcends the confines of "pure" epistemological debates, one that spills over into the messy social and political realm, one that affects how we interact and by what set of moral codes we, inside and outside the technoscientific community, abide. As he says: "Nobody fights for knowledge and truth. Everybody fights for the sake of fighting. That is knowledge itself, and the only truth." (Ibid., 48) One would not have guessed that the rejection of Rorty's tolerance in Chapter 2 and the adoption of critical, even heated, debates among critics (as shown in subsequent chapters) would lead to an incessantly vital battlefield of words and deeds. Are we doomed to fight for the sake of fighting alone? Is this our inevitable nightmare? I hope not; and by this I do not mean that I expect perpetual peace à la Kant.

Because of the rhetorical force that accompanies Serres's claims and pronouncements, it is important to shift our discussion of the epistemological concerns facing the twenty-first century into the "personal" realm of philosophers of science in particular and intellectuals in general. For I have no doubt that any sort of reductionism, whether that of the social constructivists (the Edinburgh Strong Program or any other "weaker" one), the Marxists, or the feminists, is bound to require a "reflexive" move (Bourdieu 1992), a move that in going beyond epistemological proclamation recognizes ideological commitment and metaphysical convictions or beliefs. To begin with, such a move, as some Marxists and feminists have argued all along, would illustrate that the rejection of objectivism is a rejection of the pretense of the certainty, the universality, and the permanence of all knowledge claims (all of which have become long defunct by the early twentieth century). Whatever guise the rejection of objectivism has taken, however subjective or intersubjective it has turned out, whether it calls itself Marxism, feminism, or postmodernism, this rejection opens the possibility for a more elastic appreciation of the epistemological options available for the study of reality. In their different guises, all of these options need not compete with each other, but could complement each other as ways by which to increase the imaginative content of epistemological claims.

Imaginative speculations were discarded by modern philosophers and scientists, so that when Giordano Bruno was burned at the stake in 1600, his theories of multiple universes burned up with him. It may be an exaggeration, but one could say that not until Einstein's theory took hold in the twentieth century did the scientific establishment reconsider the possibility that our galaxy is only one among many. Of course, the previous statement plays right into the hands of those distinguishing between the many-sidedness of social reality as understood, for instance, by Max Weber (Bourdieu 1990, 21) and "natural" reality or the reality of nature, a distinction whose time has passed. That is, only when "real" scientists speculate about the nature of planets is it legitimate for poets and postmodernists to speak of multiplicities and labyrinths.

Be that as it may, I would like to bring back the dialectical tension between imaginative moves or scientific conjectures and some sense of reality. No matter how differently contextualized in India or France, whether today or two thousand years ago, when a stone is thrown, are not similar gravitational forces at play? Perhaps it is at this overly simplified level that the present discussion begins. Now, what exactly that sense of reality is or means remains an open question, as the history of science illustrates in abundant and quite exciting ways. Moreover, as mentioned earlier, Eco aptly notes that there is a semiotic moment in the reception and consumption of data, even about cows: "A herd of cows is perceived in the same way by an Italian and an Indian, but for the former it signifies abundance of food, for the latter abundance of ritual occasions." (Eco 1994, 90) Finally, the discussion becomes necessarily more complex when the reality of the Holocaust is under scrutiny as opposed to that of a stone. According to Lyotard:

> But, with Auschwitz, something new has happened in history (which can only be a sign and not a fact), which is that the facts, the testimonies which bore the traces of *here's* and *now's*, the documents which indicated the sense or senses of the facts, and the names, finally the possibility of various kinds of phrases whose conjunction makes reality, all this has been destroyed as much as possible. (Lyotard 1988a, 57)

Note that Lyotard is concerned with the conjunction of phrases and statements, observation reports, if you wish, that "make reality" in the most straightforward sense. Is it the case, then, that when they are "destroyed," the reality that they "made" is "destroyed" as well? Did the concentration camp in Auschwitz not exist because it was destroyed?

These kinds of questions bring up the problematic notion of proof as it is applied to knowledge-claims about reality or nature, whether understood empirically or conventionally. For what we indeed expect of data reports is a verification such that one could prove, in principle and in practice, that what is said is true. Lakatos undertakes to deconstruct the notion of proof as it applies to mathematics so as to show that the absolute certitude associated with mathematics is bound by conventional terms, background information, and axiomatic presuppositions whose truth is assumed (Lakatos 1976).

Lyotard recommends that the historian "must venture forth by lending his or her ear to what is not presentable under the rules of knowledge," so that Auschwitz becomes "the most real of realities in this respect." (Lyotard 1988a, 57–58) The Holocaust may turn out to be a more appropriate test case for the inevitability of reality as compared to a stone or a laboratory experiment, for questions of facticity and evidence, what should and should not count have been raised by revisionist historians who claim that the liquidation of the Jews never occurred. The stakes in this case, I submit, are much higher than the disputes about low-energy conductivity.

It follows from the brief mentioning of the reality of the Holocaust that all epistemological debates have strong psychological antecedents that motivate their formulation and predetermine what political and ethical conclusions would be admissable (see also Sassower 1993a, Ch. 2). Skepticism and ambiguity arise where human intellect and emotions play the dual role of establishing physical and psychological stability while undermining it. Perhaps de Beauvoir's sense of having to win every time a contest is announced can now be better understood. Similarly, one can better appreciate Haraway's call for a socialist-feminist agenda.

Within the context of twentieth-century capitalism in the western world, the need to imagine has been magnified tremendously. What I mean here is not limited to extreme cases that may be claimed to be unrelated to modern capitalism, such as the technoscientific products of World War II, from the atomic bomb to the gas chambers in concentration camps. Instead, I mean the organization and function of the scientific community in its linkage with the industrial-state complex, wherein ambiguity is mitigated by hedging one's bets in the stock market of ideas. The imagination is considered wasteful and irrelevant, unless it is demonstrable that space shuttles can be commercially useful or that Virtual Reality can be commodified and sold to the public. Haraway's Cyborgs (Haraway 1991, esp. Ch. 8), then, have value either as science fiction (which sells comic books, novels, and films) or as cyberspace for corporate communication and the transmission of data. In either case, though, if these ideas as products are not commercially viable, they become no more important than the fragments of the preSocratics, relics of yesteryear for the amusement of idle scholars.

How can I simultaneously insist on the inevitability of reality and the necessity to imagine reality? Reality, for lack of a better term, provides the raw material with which to reconstitute, transform, and view it differently. Herein lies an excitement of the creative imagination in its acceptance of the inherent ambiguity of reality so that it lends itself to empowered reconfiguration. The reality with which we interact ranges from the vegetation we grow and harvest, the animals we come across, the mountains and the seas, the daily consumption of food and air, the clothes we wear, the mobility we enjoy, and the techniques of survival we have developed.

Perhaps the technoscientific community should concern itself not only with solving the puzzles presented to it by its leadership and the leaderships of industry and the state, but also with the dreams of neighbors and friends that all too often are dismissed as unrealistic or too prosaic. The exchange of information between technoscientists and the public should be based on pedagogical structures that nourish the improvement of the conditions under which all humans imagine, as I will argue in Chapter 7. Is it not time for us to envision a future in which happiness does not remain a term used only by poets and greeting cards manufacturers?

Postmodern Political Economy

The rest of this chapter moves from a general discussion of the need to reconceptualize theories that traditionally come under the label of political economy to a more specific analysis of corporate structure and culture in the information age. The main argument of this chapter is not limited to comprehensive proposals concerning the need to revise old-fashioned and outdated neoclassical models and principles. Instead, it tackles the difficult restructuring of corporations in a constantly changing cultural climate that defies the classical categories of capitalism and socialism. In many ways I outline an agenda for future scholars and practitioners and not a detailed blueprint with an attached budget. It is, therefore, an invitation for critical negotiation and not an injunction.

A recent attempt to reconstruct economics has been limited to the reconsideration of the role of neoclassical economics in such a reconstruction (Hodgson 1992). The main flaws of neoclassical theory are enumerated especially in terms of its lack of a moral dimension, the unrealistic portrayal of markets, the disregard for uncertainty in decision making processes, the centralizing features of the Walrasian model, the problematics of time measurement as a variable in the econometric model, and the disregard of the import of money as an end and not merely as a means of exchange. As a survey of the critical literature, this attempt to reconstruct economics brings together some well-known concerns with the neoclassical model as it has been perpetuated at least in textbooks and western capitalist culture. As the author says:

> A fundamental problem with any attempt to incorporate neoclassical theory as even a subset of a limiting case of a wider theory is that it is founded on an atomistic ontology and a mechanistic metaphor. In short, neoclassical economics is steeped in the metaphors and presuppositions of classical physics. (Hodgson 1992, 757)

The atomistic ontology refers to the old-fashioned analyses of individual preferences and the curves collected from such a set of data, as opposed to the realization that what is in effect put forth is an ideal-typical case in the Weberian sense that may or may not be generalizable for all individuals making decisions under conditions of imperfect information and disequilibrium. Moreover, mechanistic metaphors are misused in economic theory and model construction because their simple-minded causal relations defy the realities pertinent for an economic analysis.

I find this assessment correct, though certainly not novel at this point in time. For instance, Howard Sherman (1993) deploys a similar critique of neoclassical economics (and of official Marxism) in his advocacy for critical Marxism. What may be novel is the alarm sounded in this particular survey, one that comes from the less alarmist orientation of Institutionalists (as compared to the alarms already sounded by Marx and definitely repeated over the years by generations of some of his disciples). As the author claims: "To drive the point home, we may suggest that the Walrasian market model conjures up the idea of a totalitarian police state rather than a liberal market system, despite common rhetoric to the contrary." (Hodgson 1992, 754) Of course, the antecedent of this alarmist view is apparent in the critique of Karl Polanyi (1957) in which he illustrates the political conditions that must pertain in order for the so-called free marketplace to function efficiently.

Whether one wishes to proclaim the need for "pluralism" (Hodgson 1992, 761) or not remains an open question, because it is not clear if the proclamation follows Lawrence Boland's plea for methodological pluralism in the Popperian sense (Agassi 1991 and Boland 1989) or the kind of methodological anarchism advocated by Feyerabend. Perhaps any talk about pluralism becomes muddled when one conjures up images of "tolerance" and open-mindedness. It may be difficult, if not impossible, to maintain a level of pluralism and even relativism, while adhering to a set of criteria according to which economic judgments and decisions are made. And it is exactly at this point of the discussion, whether it turns into the question of a minimal state intervention in the postlibertarian discourse (Sassower & Agassi 1994) or the question of market socialism of the postmarxist discourse (Cf. Albritton 1993), that I wish to interject the lessons of postmodernism the way I understand their applicability to economic thought.

Before beginning this particular exploration, one that may be useful for anyone attempting the reconstruction of economic theory, I wish to clarify one of my basic assumptions. I am committed to the analysis and study of economic conditions and situations as the most coherent and significant means by which a cultural critique can be undertaken. That is, as far as I am concerned, Adam Smith and Karl Marx, despite the legendary portrayal of them as mortal intellectual enemies, are in fact both committed to the assumption just described. They are first and foremost concerned with economics not as a reductionist field of research that holds the ultimate answers for the questions of the day, but rather as an inevitable set of variables that must be part and parcel of any equation that tries to describe the human condition and prescribe a better future. Both are concerned with a better world, a world in which individual happiness is protected and enhanced, where justice reigns in the face of evil and bad luck, and where human relationships combine moral respect with personal well-being. Having said this, I do not mean to minimize the differences in their methods of analyses and the radical differences of their conclusions.

If political economy is to be a useful backdrop for cultural critiques (as Marxists and Institutionalists continue to demonstrate so eloquently), it should adopt a postmodern garb, one that legitimates a complexity of analyses that transcend econometric models. In this sense, then, a postmodern orientation is not a fad heralded by the avant-garde, but an admission that the debates in the history of political economy are still relevant today. Neither a high-powered hardware nor a sophisticated software will dissolve the complexity of an informed cultural critique.

In previous chapters I examined some tenets of postmodernism in light of their adherents and critics, but I wish to reiterate some of them here with the aid of a self-proclaimed critic who may be considered a postmodernist. Latour is correct in claiming that "[t]he postmoderns believe they are still modern because they accept the total division between the material and technological world on the one hand and the linguistic play of speaking subjects on the other," (Latour 1993, 61). As we have noticed earlier, this attitude is shared by the most vocal critic of science from the French postmodern quarters, Lyotard. Though arguing repeatedly that "[s]cientists, technicians, and instruments are purchased not to find truth, but to augment power" (Lyotard 1984, 46), Lyotard still distinguishes between empirical descriptions and opinionated prescriptions, as seen earlier.

The question of the privileged status of the scientific discourse is of particular interest to economists, not only because of the devastating critique of science already formulated by Marx against classical theory, and Engels's equally devastating critique (1894), but also because, as critics, political economists claim themselves to be scientists first and foremost. It is not clear whether or not the claims of economists warrant the status they crave (Sassower 1993a). So, the question of a privileged scientific discourse is still part of the appeal of economists trying to gain and maintain scientific status for their work.

I am not sure that Latour is correct in his assessment of the postmoderns, though I find his analysis of the moderns quite convincing. For example, he says that the "[m]oderns do differ from premoderns by this single trait: they refuse to conceptualize quasi-objects as such. In their eyes, hybrids present the horror that must be avoided at all costs by a ceaseless, even maniacal purification." (Latour 1993, 112) And, unlike the moderns, I do believe that the postmoderns are open to the thinking and appreciation of the so-called premodern, at least in the sense described earlier in this chapter in relation to preSocratic epistemology. This, incidentally, would be of great interest to potential economists who should admit that the phenomena they study and predict are hybrid phenomena: human and natural and discursive, all at once.

But still, how does Latour define postmodernity? Is it only in terms of modernity? Can this historically relativized definition be avoided, as Lyotard prefers? In Latour's words: "The postmoderns have sensed the crisis of the moderns and attempted to overcome it . . . . Take away from the postmoderns their illusions about the moderns, and their vices become virtues—nonmodern virtues!" (Ibid., 134)

Defining and fastening one's radar on a particular strand of postmodernism is a slippery slope (Albritton 1993, 25–26), for no postmodern critic will readily align her or his critique with another's. As a group, these critics are perhaps more individualistic than any other group of scholars, including the champions of individualism among the mainstream neoclassical economists. Perhaps they remind us of a convention of Marxists or Institutionalists whose members berate each other for not being Marxist or Institutionalist enough, or for being *too* Marxist or Institutionalist, respectively. The simple fact that they are all Marxists or Institutionalists (as the case may be) recedes into the background and is overlooked rather than taken for granted.

Latour concludes his assessment of postmodernism in the following way:

> As always, however, postmodernism is a symptom, not a solution. The postmoderns retain the modern framework but disperse the elements that the modernizers grouped together in a well-ordered cluster. The postmoderns are right about the dispersion; every contemporary assembly is polytemporal. But they are wrong to retain the framework and to keep on believing in the requirement of continual novelty that modernism demanded. By mixing elements of the past together in the form of collages and citations, the postmoderns recognize to what extent these citations are truly outdated. Moreover, it is because they are outmoded that the postmoderns dig them up, in order to shock the former "modernist" avant-gardes who no longer know at what altar to worship. (Latour 1993, 74)

So it should be clear that I am not advocating the wholesale adoption of postmodernism. Instead, I wish to inject a postmodern note into economics *qua* science so it becomes premodern political economy. How is this to be done?

First, in the postmodern spirit of recalling the insights of premoderns who accepted ontological hybrids and the polytemporality of all constructed models and epistemological judgments, I implore economists to think of themselves as political economists. As political economists they would have to be familiar with the Marxist and Institutional traditions just as much as those of classical and neoclassical economics. This proposal, then, is not limited to the practices of economists and the introduction of additional variables to their models, econometric or not. Instead, this proposal directs the changing of the curriculum in the undergraduate and graduate programs so that the context for economic analysis would become an integral part of every student's education. I say this not as a mere recommendation for those interested in a broader scope of their education, but as a necessity for any political economist who wishes to understand, predict, or change the economic conditions of any society. (I will return to education in the next chapter.)

Second, and related to the first proposal, while waiting for new generations of political economists, it is imperative that contemporary economic analysis should incorporate the insights of sociologists, psychologists, philosophers, and many others in and out of the academy. The artificiality of disciplinary boundaries has been recognized recently by those developing interdisciplinary programs across the academy, from Biochemistry to Cultural Studies. In the meantime, though, it will be useful for economists to invite non-economists to their conventions and classrooms, to their conferences and journals. The invitation is not only for the sake of receiving the conclusions of other intellectuals, but also for the sake of a critical evaluation of the work of political economists themselves. In an era that cannot rely on self-policing even within the scientific community (e.g., latest revelations concerning breast cancer research), it would be extremely useful to "import" critics at low costs and with a potential for extremely high returns. In addition, political economists should follow Haraway's lead in laying out their moral and political commitments as variables whose relevance is foreshadowed. They can find a model for this proposal in the works of Gunnar Myrdal (1969).

Third, if the material conditions of a society are important determinants of the well-being and potential growth and progress of that society, as economists on the "left" and "right" side of the political spectrum agree, then their structural change and development must remain the basis for any analysis and application whatsoever. Though claiming to live in a poststructuralist world, I am bound to remain a structuralist through and through. Perhaps the confusion of poststructuralism is to conflate the need for structural analysis with a functionalist reduction. It is a mistake to demand of social scientists only a functionalist analysis, for there are situations that do not lend themselves to a functionalist framework at all. But to suspend functionalist conclusions is quite different from rejecting the reasonable expectation that explanatory models of causal relations, for example, benefit dramatically from highlighting the structural conditions that pertain in particular situations (See the SSC in Chapter 1).

Fourth, I propose that all political economists temper their studies and the conclusions that emanate from them with a heavy dose of rational skepticism and relativized rationality in the Popperian sense (Cf. Agassi & Jarvie 1987). Instead of assuming that the scientific jargon can and should shield their work from criticism, they should welcome the critique of others by making explicit the limitations of their particular analyses, as well as the general theoretical framework within which they are undertaken. A degree of humility would ensure a degree of honesty and perhaps even a degree of intellectual integrity so sorely missed in the academy and definitely outside it. Perhaps, unlike the declarations of other theoreticians, those made by economists are often taken literally as true by policy makers whose decisions affect whole populations. For example, models in developmental economics are admittedly custom-tailored to each country. It would make sense if the curriculum reflected this situation, and thereby demanded a more open-minded orientation by brilliant econometricians.

To say, as so many have in the past twenty years, that the economy of the United States has shifted from manufacturing to service is to admit that by the end of the twentieth century, the economy will have shifted its emphasis to developing technological methods. The incorporation of these methods into existing industries, and the way they bring about the inventions of new technologies and goods, are models different from the previous cornerstones of American capitalism. My proposals for change are less radical than might appear at first, for the free enterprise system has been overshadowed, if not eliminated, through direct government subsidies, tax incentives, and legal protection and constraints (e.g., patent and copyright laws).

If, despite anti-trust laws, there is no pretense anymore to compete domestically (as we observe cartel-like behavior or price-fixing by the airline industry, for example), it still remains an open question whether or not the United States would like to compete internationally. As the residues of the North American Free Trade Agreement debate are becoming apparent, the United States may have no choice but to join the international marketplace as a partner and not as a competitor. If this turns out to be the case, it is even more crucial that the economy be restructured to deal more efficiently and thoughtfully so as to become a cooperative environment to develop new methods of technological innovation.

One way to explain more concretely the issues at hand is to analyze the computer industry. This is an industry composed of a variety of sub-industries, some devoted to the development of hardware, such as supercomputers of the Cray kind, some devoted to the development of software, such as Microsoft's "Windows," others devoted to computer chips that blur the distinction between hardware and software, and still others to the conceptualization, organization, retention, and dissemination of databases. What is increasingly clear is that what the computer industry "sells" is not limited to products, such as personal computers and spreadsheets, but instead includes the methods by which new technologies "deal" with information. Who is in control of these methods? Who has the right to sell them? In whose interests are they developed, sold, and made obsolete through design changes?

When these questions are raised we are at once retreating historically as well as proceeding (as suggested above) to an era in which it is incumbent on us to speak of political economy, a locus of social, financial, political, legal, and moral concerns that are intimately connected to each other and that inevitably affect our daily life. Should the multibillionaires, like Bill Gates of Microsoft, who own and run megacompanies be in charge of our "life" in the next century? Should such an individual be in control of the methods by which the rest of the computer industry (and eventually the rest of the economy as well) conducts its affairs? Incidentally, I do not wish to develop a conspiracy theory concerning the activities of multibillionaires; rather, I wish to illustrate the elasticity of the capitalist system that need not induce its leading members to compete in the traditional sense, but may show them how to consolidate their power and longevity by buying out all competing ideas.

The legal protection enjoyed by the Microsofts of this country are such that their longevity (not to mention their survival) is guaranteed regardless of the social costs borne by society as a whole or the benefits enjoyed by any segment of that society. This is true both in terms of direct tax incentives and in terms of government financed research that becomes available as indirect generous subsidies to private corporations. This situation is exacerbated by the following quandary: while copyright and patent laws, regardless of their variations, protect products from any and all competition (i.e., duplication) for anywhere between fourteen and eighteen years (with renewal options in most cases), the competitive "life" of a computer-related product lasts only three to five years at the most. Should these laws be changed? Should they be fully enforced, despite their short-lived usefulness, or relaxed, if not completely abolished? If they were to be relaxed, would there remain enough incentive for private corporations to invest in product research and development?

These questions are not theoretical in nature, for they supersede the general concern with intellectual property and academic freedom (e.g., Buskirk 1992, Cordes 1993). With a stroke of a pen (and the aid of a thousand technicians) someone like Microsoft's Gates can turn a thriving technological component of a research program into obsolescence, thereby increasing the stakes for those interested in encroaching on his market share. No matter what antitrust laws are in effect, and no matter what the Contestable Markets Theory claims (Sassower 1988), individual corporations can, and in fact do, dictate the future of their industry. Now, of course, this description is much more complex, for there are reasons for not enforcing patent rights for some products, if there are financial benefits from public involvement in the use of other products related to the original patent. But I will not pursue the complicated legal and financial details of such situations. In general, the concentration of power in few hands is disturbing, not because the free enterprise ideology is undermined, but because the unfair practice of these corporations is legitimated by this ideology.

Coordinated Elasticity: The Virtual Corporation

It makes sense to shift away from the pretense of a free enterprise ideology because this ideology is untenable at the end of the twentieth century. Perhaps we should avoid the standard classifications of political economy, such as capitalism, socialism, and anarchism, and adopt the notion of *coordinated elasticity*. I define coordinated elasticity as a constellation of material conditions, the use of which is coordinated by a variety of institutional representatives so as to have the widest public benefit, while succumbing to some of the principles of efficiency (as the minimization of the detrimental secondary effects of traditional business cycles). This term incorporates the postcapitalist need to collaborate in an age of increasing research and development costs so that unnecessary duplications can be avoided, while allowing sufficient flexibility for innovations and counterintuitive experimentations. Put differently, there is always need for the repeatability of technoscientific experiments; yet, there is no need to begin all research from scratch, constructing additional laboratories where there is ample opportunity to use existing ones.

The notion of coordination is not that of a centralized authority nor the one associated with the "perfect information" of marketplace prices that emerge under conditions of supply and demand. Instead, it is a way to minimize the devastating effects of business cycles when oversupply or overdemand are accompanied with periods of hyperinflation or high levels of unemployment. The notion of elasticity has to do with rethinking neoclassical microeconomics that focuses on the firm. The firm may turn out to be without walls, so to speak, for flexible schedules at home and computer networks and modems would hasten the elimination of corporate headquarters and transportation costs, not to mention pollution.

What legal conditions must be set in place on behalf of noncompetitive cooperation? It would seem, for instance, that the enforcement of revised copyright and patent laws should provide protection to a "product" and its "owners" only if the expense of its development is borne by all the collaborators *and* if all have equal access (or rights *qua* owners) to its use, albeit only for a much shorter period of time than currently in practice. But in order to explicate the need for a changed legal framework, that is, a framework that will more properly and adequately reflect a changed ideology and culture, perhaps it would be useful to examine the possibility of noncompetitive cooperation. (I mentioned the need for collaboration and the implementation of user fees when I discussed the Superconducting Supercollider in Chapter 1.)

The proposal to shift toward a noncompetitive marketplace of ideas and business practices is not the brainchild of theoreticians bored with the neoclassical ideology of the past century. Instead, it is a necessity. It is necessary to share knowledge and create institutional partnerships that transcend intellectual fascination and "one-upmanship" and move toward the honest disclosure of the limits to anyone's capacity to know "everything" (Cf., Faust 1984). One form that illustrates our concern has been championed already under the name of the Virtual Corporation: "It's a temporary network of companies that come together quickly to exploit fast-changing opportunities. In a Virtual Corporation, companies can share costs, skills, and access to global markets, with each partner contributing what it's best at." (Byrne 1993, 98) Incidentally, the term "virtual" did not originate from the phrase "virtual reality" but rather from the computer industry that talked about "virtual memory" to "describe a way of making a computer act as if it had more storage capacity than it really possessed." (Ibid., 103) The main characteristics of this new corporate model are enumerated under the labels of excellence, technology, opportunism, trust, and no borders. In short, the virtual corporation tries to expand the capacity of existing corporations by creating complementary corporations that add new dimensions to already existing activities.

In old models of collaboration opportunism motivates the sharing of information. Shared information is minimal and is only shared for the sake of taking advantage of a third party, such as consumers (cartel-like behavior in macroeconomic theory). If anything is accomplished, as for example the pricing practices of OPEC, it is despite incomplete information and lack of trust. The new model distinguishes itself because it requires that every partner put up its best personnel and its latest technology, not the kind of material that it no longer needs for exclusive use in order to have a competitive edge. This, of course, requires a great deal of mutual trust, and a sense that shared costs and risks would enhance not only current productivity but also future growth.

It seems reasonable to follow this line of thought because of what I consider to be three sets of perceptions about the material conditions of the United States: first, an inherent scarcity of funding resources, no matter how much accumulated capital there is, because all research and development projects are by definition too risky for most individual entrepreneurs; second, an inherent scarcity of creative talent, because most educational institutions undermine personal creativity in their efforts to indoctrinate all students and researchers under the pretense of "initiation rites" (e.g., Fleck 1979, 54); and third, a scientific precondition required for the sake of legitimacy: multiple verification or confirmation of experimental results. With these three sets of perceptions of the conditions under which coordinated elasticity may operate, one seems to be facing an insurmountable obstacle for any traditional technoscientific research program. It seems that a reasonable way to overcome or directly deal with these perceptions (and thereby reconstitute conditions for change) is to spread the costs and risks to as many willing partners, or else seek the benevolent support of a government agency.

If so-called basic research is understood to benefit the entire society equally and therefore should be funded by the government, and if the application of that same research is supposed to be carried out by private corporations for their own profitable benefit (see for example NASA and the General Information System), then we are torn between a model of collaboration on the one hand and competition on the other. Furthermore, undue advantage is gained by private enterprises at the expense of taxpayers without any return to these same taxpayers (who are in fact unsuspecting investors in these enterprises). How does one balance the two, and still maintain a coherent ideology? It is here that I wish to introduce the twin concepts of relativized rationality (from Chapter 5) and coordinated elasticity. The virtual corporation illustrates that the rational choices individuals and corporate executives make are not absolute, but context-specific. It therefore makes no sense to repeat the standard views of "rational economic man" (Hollis & Nell 1975). Similarly, to speak of coordinated planning in the Soviet sense of a five-year plan makes little sense. In their stead, it is incumbent on us to reconsider old terms and concepts, revise them, and offer more appropriate terminology and grammar for the next century.

To offer a revised vocabulary for the next century means to offer a revised curriculum (as I will further argue in the next and final chapter). In order to revise the curriculum we need to convince instructors that they ought to write off their intellectual investment of yesteryear and accept the challenges of new risky investments whose returns remain dubious. So, the difficult task of reorientation, whether called postmodern or not, rests much more on the shoulders of established academics than on the shoulders of students. For it is they who must prepare and entreat a new generation of students to dream of different ways of thinking and doing.

One such dream has been the Superconducting Supercollider, a dream that failed to come true (Chapter 1). Perhaps it is presumptuous to argue that if advocates for this technoscientific project had used the postmodern orientation to political economy they would have achieved collaborative goals in a coordinated and elastic fashion, both national and international. Yet, as a case study, the SSC illustrates why and how it is necessary to transform the rhetorical devices of the technoscientific community as used since World War II and avoid the vocabulary of capitalism and socialism alike.

Chapter 7

Art and Imagination: Intellectual Responsibility

Cultural Context

Why juxtapose art and imagination to a postmodern discussion of technoscience, political economy, and the Superconducting Supercollider? There are several reasons, some trivial and transparent, some more convoluted and complex. For example, the enigma of the visual representation, the pictorial artifact, is that at first glance it seems transparent and immediate, clear-cut and worth one thousand words (to quote Neurath)—but it also brings together a multitude of variables, some social, some semiotic, and some cognitive, so that the complexity of composition and deconstruction undermines the appearance of transparency (incidentally, similar to the concerns of translation, as mentioned in Chapter 3).

A postmodern orientation can make use of art and imagination as a means of providing alternative interpretations and modes of expression. It is not only that postmodernity was originally aligned with architecture and only later was deployed as a mode of critique of technoscience. Rather, we should revisit these artistically informed origins in order to recall the inspiration associated with the postmodern move. If the postmodern move in intellectual life turns into jargon-ridden posturing, then its value will be limited to degree-hungry academics, and perhaps some architects and artists.

Second, there is a gap to which Eco alerts us between comprehension and verbalization:

> Fabbri points out that an enormous gap separates *comprehension* and *verbalization* and that the conflation of the two derives from the myth of the word (which also dominates semiology), whereby something has meaning only when it can be verbalized, translated into words and *thought*. (Eco 1994, 92)

In this respect, the question is no longer whether an artistic rendition is transparent or opaque, clear-cut or complex, but rather even if it is transparent and clear-cut, must it be verbalized to be comprehended? Contemporary society relies heavily on verbalization, the words and sentences we utter and listen to, whether written or on camera. When we are awestruck by a pictorial representation or an artifact, do we have to put its "meaning" into words? Do we owe it to the experience to translate it into words? Do we owe it to our fellow humans to do so? At times the sense of awe is at the heart of technoscience, and at times this sense must be articulated in order to present or re-present it to funding agencies.

Perhaps it is time to infuse technoscience with the age-old adage of the art world: art for art's sake. Now, there are numerous arguments on behalf of and against this adage. And of course there are compelling reasons for adopting instrumental-rationality in the face of scarce resources. But here I would like to propose two questions: first, is it indeed the case that *all* resources are scarce? (and how does one define scarcity and resources? see Sassower 1990a) And second, can we afford not to support technoscience for technoscience's sake? (Or for that matter, art for art's sake?) So, the reason for bringing up art and imagination when discussing postmodern technoscience has to do with an analogy drawn between these "two cultures," contrary to C. P. Snow's famous view of their discordance (1964), concerning human creativity, inspiration, communication, and survival.

Third, the juxtaposition of postmodern technoscience and art and imagination brings into play the relation between political economy (capitalist mode of production or any other) and art. Lyotard has the following to say in this context:

> Yet there is a kind of collusion between capital and avant-garde. The force of scepticism and even of destruction that capitalism has brought into play, and that Marx never ceased analyzing and identifying, in some way encourages among artists a mistrust of established rules and willingness to experiment with means of expression, with styles, with ever-new materials. There is something of the sublime in capitalist economy. It is not academic, it is not physiocratic, it admits of no nature. It is, in a sense, an economy regulated by an Idea—infinite wealth or power. It does not manage to present any example from reality to verify this Idea. In making science subordinate to itself through technologies, especially those of language, it only succeeds, on the contrary, in making reality increasingly ungraspable, subject to doubt, unsteady. (Lyotard 1991, 105)

It seems puzzling to find a critic of capitalism discover in its manifestation and expression, in its motivation and incessant reproduction, something of the sublime. The sublime, whether Kant's or Lyotard's, is associated with art and imagination, creativity and inspiration, the unspeakable and unpresentable. Capitalist production, distribution, and consumption are all materialist activities whose nature seems debased and frowned upon: these are activities one must perform in order to gain enough leisure time to enjoy the sublime. But Lyotard points out the paradox of the avant-garde art world: capitalism, against which it rails routinely, is also the means by which it feeds its needs and aspirations.

If this is the case in art, why is it not also the case in technoscience? Every project, the Superconducting Supercollider included (see Chapter 1), relies heavily on the support of capitalism while defying its mentality (or at least its corporate image). It is not that technoscientists, like their artist counterparts, refuse to make money. Rather, they too refuse to abide by the rules and constraints of the corporate world. Technoscientists probably want the same autonomy craved by artists; they wish to break new ground and be their field's avant-garde. Their idiosyncrasies are touted and celebrated as a form of nonconformity, and their status is contested in a fashion that differs little from that of artists, especially since it is undertaken against the backdrop of capitalist socioeconomic stratification.

Now, I may be accused of focusing on small and even atypical segments of the artistic and the technoscientific worlds. Fair enough. Yet, these small segments have a symbolic power and cachet well beyond their size and regardless of how typical they may be deemed. Moreover, these small groups of individuals inspire generations of students to try to emulate them, though they will never be able to do so (either because they happen to be less fortunate, less talented, or less driven). They turn out to be role models in the best sense of the term. There is, though, a difference between artists and technoscientists: artists have an image more public than that of technoscientists. The license given to artists—to be "crazy," to behave "inappropriately," or to be different—should be extended to technoscientists if we are to hope that some of them will push themselves beyond the confines of technical prowess and puzzle solving bravado (à la Kuhn).

Fourth, in the context of postmodern technoscience, in the age of Haraway's "cyborgs," the imagination struggles to grasp and keep up with the realities and phenomena that surrounds it. It is no different from the kind of advice given by Lyotard concerning the unsteady nature of reality or from the warning given by Gunther Anders (after Auschwitz and Hiroshima) concerning the need to imagine ahead of our technoscientific capabilities (Eatherly & Anders 1989). When the imagination of science fiction authors preceded the actualization of certain devices, such as the video telephone shown in *Metropolis* in 1927, it was unclear if the ideas authors and artists had were designed to improve the human condition or bring about its final destruction (as in the gas chambers of Nazi concentration camps).

When we say that something is unbelievable, what do we mean? Is it that the phenomenon was not imagined until it took place? Is it that certain conditions brought about an accidental constellation of circumstances that was unimagined until it actually happened? Or does it mean that we have failed to nurture the imagination enough or at all? Perhaps these questions hinge on the Popperian resolve to propose conjectures that will be falsified as opposed to the comforting reliance on inductive knowledge, that is, knowledge that relies on past experiences.

Fifth, the question about nurturing the imagination can be understood on different levels. For some, it is an issue of paying attention to one's imagination, the fanciful, the temporary insight one comes across. To others, it is a deliberate activity undertaken by artists for the sake of art, the enrichment of their souls, and the entertainment of others. And for still others, it is a question of education, the socialization of as many individuals as possible to retain their childhood awe of the world and their immediate surroundings, to explore new forms of expression, to break the rules and *status quo*, and to learn how to cultivate creativity. On the whole, this is a philosophical question that goes to the roots of the ideological backdrop against which societies mandate, manage, and legitimate their educational systems.

It is unclear whether the imagination can be cultivated at all. Some would argue that either one does or does not have an active imagination. If one does, then it is assumed that no cultivation is needed. If one does not, then it is likewise assumed that no cultivation will make any difference. I would like to propose the contrary: whether one does or does not have an allegedly active imagination, it is crucial to cultivate whatever one has. If indeed there are no results, so be it. But by the same token, a bit of cultivation may help one's imagination soar to the ends of the universe. One may not know it is there to be cultivated, so any outlet may make a difference. Moreover, technoscientific innovations, just like artistic ones, activate the imagination and transcend the confines of its own history—if anything happens at all in our surroundings, we have to keep up with it; we have no choice in the matter.

Sixth, educational systems that are aware of the imagination and are interested in cultivating it may do so for the sake of enriching humans' souls or creating an environment that is more hospitable to human communication than that based on instrumental rationality, in the sense of the one-dimensional man about whom Herbert Marcuse talks (1964). However, it may turn out that the use of the imagination in postmodern technoscience is not a "luxury" or an exogenous variable: it is not about feeling better about oneself and one's surroundings, but about alternative forms of comprehension. There are practical reasons why the imagination ought to be cultivated. It is with this "ought" in mind that I turn to the next section.

Conscience and Responsibility

In the course of his correspondence with Claude Eatherly, the pilot who dropped the atomic bomb on Hiroshima, Gunther Anders, a Viennese philosopher, sends his "Commandments in the Atomic Age" originally published in 1957. In it, Anders explains that after World War II humanity has to reconceive its convictions and behavior patterns, "for in the course of the technical age the classical relation between imagination and action has reversed itself." He continues:

> While our ancestors had considered it a truism that imagination exceeds and surpasses reality, to-day the capacity of our imagination (and that of our feeling and responsibility) cannot compete with that of our *praxis*. As a matter of fact, our imagination is unable to grasp the effect of that which we are producing. (Eatherly & Anders 1989, 12)

Postmodern technoscience, as understood here, exemplifies Anders's concern over the relationship between imagination and action. If the traditional transcendence of the imagination over mere human action was the acceptable model, then it is an outdated model. The model of the technical age, as Anders calls it, requires that the imagination play catch-up to the actions of humans. Human imagination lags behind human action, and the call for the imagination to transcend action is a warning call that we need the imagination to save lives, to survive, perhaps in the sense concerning translation that I mentioned in the Preface.

American politicians and military leaders failed to imagine all the secondary effects of dropping atomic bombs on Hiroshima and Nagasaki. For them it was a strategic decision, one having to do with the military threat of Japan; it had to do with a way of averting too many casualties, of ending the war; it was, in short, not about the atomic bomb *per se*. Their imagination was confined to a human balance sheet, an equation written on some utilitarian drawing board, one showing that more lives will be saved than lost if a bomb of such a magnitude was dropped on a civilian center in Japan. But how could the imagination of political and military leaders be about anything but the horror of the atomic bomb? They chose that bomb over the conventional bombs at their disposal. Did they not imagine the different level of destruction that atomic explosion and radiation would bring about? Of course they did.

Yet, while they calculated the results of atomic damage, they did not calculate the results of dropping conventional bombs on the railway system that led to concentration camps in Europe, or if they did, they decided against bombing. Why? Historians can provide a variety of answers, from sheer ignorance on the part of American leaders to insidious racism and anti-Semitism. Instead of following their alternative answers, perhaps I can use Anders's concern with the imagination and say that the only way to justify the application of strategic calculations during World War II is that in the case of Nazi concentration camps, Americans could not have imagined the reality of mass murder.

So human imagination must be used and cultivated, must be foregrounded to ensure that the human decision-making process not only keeps up with human actions, but perhaps even averts or prevents the most horrifying of them from taking place at all. In one of Anders's letters to Eatherly, there is an injunction for moral responsibility that rests with all of us, not only those directly involved in decision-making or in explaining the results of military actions. In his words:

> As we have to count on real creatures and not on ideal figures, it is our task to produce a situation in which even people without burning moral passion or without imagination behave as if they did it for moral reasons or out of love for mankind and all living being. (Ibid., 23)

Why focus on the correspondence between Anders and Eatherly at all? The most compelling reason for me has to do with the manner in which Anders contextualizes the dropping of the atomic bombs during World War II. The people involved in the war, those we wish to portray as guilty, are not working in isolation, nor can their actions simply be defined in terms of guilt or innocence. Perhaps unlike the guilt of Nazi leaders and soldiers, the guilt associated with the atomic age has to do with the age of technoscience, an age confused by its very success. This is an age underdetermined morally, socially, and politically because it has been overdetermined scientifically, technically, and economically. If we are to survive the age of postmodern technoscience, we need art and imagination, education, and responsibility. Only with these components in hand can we expect not to repeat the horrors of this century. As Anders says:

> As a matter of fact I think that Eatherly's case, although unprecedented, should not be regarded as isolated or unique, but rather as a prophetic example which indicates to us how man in the technical age will be bound to react to his being entangled in actions which, in the most ambiguous way, are his and not his; to actions which make him *guiltlessly guilty*. (Ibid., 65)

It is fair to interpret this last statement as one that does not sanction the abrogation of one's responsibility. Eatherly is guilty; yet his guilt must be understood in terms of his role as a pilot in the American Air Force during a war and in terms of the invention and deployment of the atomic bomb. But we have had pilots before, and we have dropped bombs on civilians before. The rationale has remained the same: by inflicting great and immediate pain on a few, many will be saved, whether on civilian or military personnel. This was the case made on behalf of the atomic bombs. What makes this case different? Is it the sheer magnitude of destruction and human suffering? Is it the use of nuclear armament as a deterrent to any enemy whatsoever, a deterrent more powerful than anything experienced in any previous war? The answers to these questions seem to require a different set of criteria and even a new vocabulary. However, except for the potential for the destruction of the entire globe, is it indeed clear that Anders's analysis must differ radically from any previous one? What is it about Anders's concern that carries us beyond the Eatherly case into the next century, a century that will be characterized by postmodern technoscience?

Art and the Monitor

When technoscientists are in the laboratory and discuss their findings, what reference point do they have? At what "reality" do they point their fingers? They obviously cannot see some of the phenomena that they observe or with which they experiment in any traditional sense. Theirs is an indirect observation that for its very existence or validity depends on indirect methods or tools of observation, devices with which the unseen is seen, with which depth perception is brought to the surface of a monitor or an inscribed piece of paper. When pictorial representations appear on the monitor or screen of a bomber travelling at the speed of sound, what do they represent? Do we in fact see anything or do we imagine that we see something? How much do cultural parameters and previous experiences overdetermine what is reported as an observation?

On the question of representation, or as some write it, *re-presentation* or even *re-present-ation*, it is necessary to acknowledge the problematic situation in which time zones are shifted, where the past is recorded in the present for future use. What is recorded and how it is recorded are questions that complicate the answers modernism and realism have provided over the years, for they bring into play variables that were considered irrelevant in classical accounts of empirical reports. Working from a Marxist tradition and the framework of the construction of knowledge (whether understood as social epistemology or as other forms of critical theory), Linda Hutcheon argues from the perspective of a postmodern feminist. I follow her line of argument, instead of reviewing the vast literature that constitutes her and my own background, for it comes closer to my concern with the conjunction of postmodernism and technoscience. Hutcheon says that:

> What postmodernism does is to denaturalize both realism's transparency and modernism's reflexive response, while retaining (in its typical complicitously critical way) the historically attested power of both. This is the ambivalent politics of postmodern representation. (Hutcheon 1989, 34)

While politicizing representation, as effects and sources of culture and ideology (Ibid., 6–7), Hutcheon insists on the paradoxical situation in which she, as a postmodernist and feminist, finds herself, one of complicity and critique which forces her respectively into inscribing and subverting conventions (Ibid., 11). Her admission parallels that of Lyotard in relation to capitalism: that which we attempt to critically deconstruct is also that without which we cannot live. This, however, does not mean that art and imagination are politically neutral; on the contrary, this foregrounds the political features and underlying principles and convictions according to which art and imagination find inspiration and audiences. But how does one become sensitized to this mode of thinking?

The training of artists on computer terminals incorporates the technical prowess required to produce the work while attending to the artists' creativity, imagination, and inspiration. The one cannot flourish or even survive without the others. By contrast, it seems that scientific training defers inspiration to either a promissory later date or to leisure time as relief from the seriousness of the work at hand. My colleague's experience with graduate students who work in computer art shows that only a few have been trained all along to simultaneously pursue and nurture both the technical and artistic sides of their work. So students are either trained as artists or as technoscientists. A student will be able to make up lost time in technical training and be better prepared to do computer art, as compared to those technically-prepared students who have a harder time "catching up" to the level of artistic prowess brought into the fray by trained artists.

This whole setup, of course, runs contrary to many popular prejudices perpetuated in the academy. For example, some science programs for premedical or engineering students argue that it is necessary to train scientists in science to the exclusion of all other intellectual distractions (Sassower 1990b). It is incumbent on the academy not to "waste" time in the education of scientists—it is the nuts and bolts they (students) have to master; and only afterwards, if they have time and energy, can scientists indulge in tertiary activities that are not directly relevant for their professional pursuit. Not only that, it is suggested that on a scale of difficulty, science ranks much higher than art (this prejudice dates back at least to August Comte's hierarchy of the sciences). Anyone can go to a museum of art or a gallery and "enjoy" art; but few people can go to a laboratory and benefit from their exposure to scientific data. Hence, art is easy and transparent while technoscience is opaque and requires rigorous training. Is this true?

I think not. It appears that the complexity of technoscientific training differs little from that of art. We must learn to "read" a painting as we do an equation: both have simplicity and elegance, beauty and complexity, and some explanatory or predictive power that matters to our very being. At the same time, both have a certain set of standards against which they are measured. Whether one is an artist or a technoscientist, one needs to know one's trade, to be engaged and committed, to be devoted and inspired. The question, therefore, concerns the conditions that bring this commitment about and not the specific results each activity can clarify for itself. In this sense, then, being an artist or a technoscientist does not matter (as we have seen above). What matters is whether, following Lyotard's postmodern injunction, one is willing to experiment, break rules, invent, and reconceive what the future may be like.

While capitalism incites us to become innovative, be part of the avant-garde, and create new horizons, it also ties us into certain social roles and economic expectations. We expect to find "starving artists" among us, whose suffering is supposed to be inspirational. By the same token, we expect technoscientists to have the latest technology with which to accomplish their work. Put differently, we expect technoscientists to be funded well enough to be able to produce anything they wish, while we expect artists to have little or nothing and still produce profound works of art. Why? Is it because a laboratory with a "clean room" is expensive while a warehouse-like studio costs very little? Is it because paint and plaster are cheap, relative to laser equipment? Perhaps computer art breaks down this historically informed dichotomy, because the equipment for computer art can be as expensive as the scientific tools for laboratory work, whose hardware and software are bound to place restrictions on what can be accomplished and what is expected. Perhaps this is a mode of thinking and doing that defies the age-old categories of art and science where the former is spiritual and imaginative and the latter is bound by equipment and mastery of a technical apparatus.

The Multiversity

In order to defy traditional categories and classifications, one should focus on venues of action and strategies, the effects of which will lead to different, if not new, categories and classifications. Because of the legacies of the enlightenments of the eighteenth century it is expected that education will take care of this transformation of consciousness. This expectation is particularly borne out today, because we recognize in the age of sociological and ethnographic self-reflexivity that the process of socialization is most profoundly exercised in school. Modes of socialization range from unabashed indoctrination to skill acquisition, and they begin as early as first grade and continue till a terminal degree is acquired in institutions of higher learning (Stanley 1978, 188, 212–213). Are we all indeed bound by the socialization processes that impose their histories on us as naive and undetecting students?

Here is Nietzsche's view as it applies to the notion of translation as a mode of socialization that needs to be historically informed:

> *Translation*—The degree of the historical sense of any age may be inferred from the manner in which this age makes *translations* and tries to absorb former ages and books .... How deliberately and recklessly they brushed the dust off the wings of the butterfly that is called moment! ... They did not know the delights of the historical sense; what was past and alien was an embarrassment for them; and being Romans, they saw it as an incentive for a Roman conquest. Indeed, translation was a form of conquest. Not only did one omit what was historical; one also added allusions to the present and, above all, struck out the name of the poet and replaced it with one's own—not with any sense of theft but with the very best conscience of the *imperium Romanum*. (Nietzsche 1974, #83)

Nietzsche criticizes those who do not translate the past into their own historical moment not as conquerors, but as those who have learned from the past and incorporated its lessons into their education. Socialization that assumes the goal of conquest is bound to miss the delights of history, the fruits of revisiting the origins of ideas, the genealogy of every conception, the process of education itself. How postmodern, then!

The Nietzschean narrative in its current permutations turns out to be quite problematic within the academic context, because it is not clear what is meant by the term "the academy." I agree with Steven Connor, who argues that "contemporary conditions present the academy with a crisis of self-definition." (Connor 1989, 19) He finds that a great deal of the responsibility for these conditions, both material and intellectual, emanate from a vision of "cultural 'heterotopia'" whose formulation is inspired by "the postmodernism debate." The dialectical tension that ensues within or in relation to the academy "allows us to see the academy as formulating, concentrating and distributing powers at the same time as it is losing them." (Ibid.) In the postmodern age of technoscience, then, questions of power relations and the matrices within which the academy operates are raised anew.

Clark Kerr's notion of the "multiversity" of the 1960s is postmodern and technoscientific in the best and worst of configurations. It deploys a strong sense of history in order to serve the changing cultural needs of its moment: it has to be many things to many people. It is powerful and marginal at the same time, providing leadership and foresight, while being subservient to the needs of the public. In Kerr's words:

> A university anywhere can aim no higher than to be as British as possible for the sake of the undergraduates, as German as possible for the sake of the graduates and the research personnel, as American as possible for the sake of the public at large—and as confused as possible for the sake of the preservation of the whole uneasy balance. (Kerr 1963, 18)

Following Fritz Machlup's concern with "knowledge production" and the significance of this economic phenomenon in the second half of the twentieth century (1962), Kerr explains what he perceives to be the importance of the production of knowledge in the university system:

> Knowledge has certainly never in history been so central to the conduct of an entire society. What the railroads did for the second half of the last century and the automobile for the first half of this century may be done for the second half of this century by the knowledge industry: that is, to serve as the focal point for national growth. And the university is at the center of the knowledge process. (Kerr 1963, 88)

It is possible to extrapolate from Kerr's notion of the knowledge industry that even if it were limited to specific modes of data gathering and dissemination, it should be pliable enough to appeal to a variety of consumers. This means that when potential consumers cannot be consulted ahead of time to modify modes of delivery of knowledge-claims, the producers have to anticipate their customers' preferences, and allow for later revisions and restructuring. How will they accomplish this task? By using art and imagination, envisioning what may be liked or disliked, found useful or useless. The knowledge industry, then, is an industry the demands on which are no different from those who produce other goods and services, because one cannot assume that supply will bring about its own demand.

If art and imagination are necessary for the knowledge industry of the postmodern technoscientific age, does the multiversity—as its home—appropriately suit its needs and character? Kerr insists that if the knowledge industry is to flourish and provide the services expected of it, creativity has to be more openly and consciously supported:

> In the arts the universities have been more hospitable to the historian and the critic than to the creator; he has found his havens elsewhere. Yet it is the creativity of science that has given science its prestige in the uni-

> versity . . . the universities need to find ways also to accommodate pure creative effort if they are to have places on stage as well as in the wings and in the audience in the great drama of cultural growth now playing on the American stage. (Ibid., 112–113)

The great drama of cultural growth playing on the American stage gives the university an incentive to participate, to engage itself in the present in order to remain a participant in the future as a multiversity, that is, as an instrument of the knowledge industry, its funding agencies, and its private donors. Whether there is drama, melodrama, or comic relief playing on stage is unclear; that something is playing is clear enough; so the multiversity can help determine the nature of the play, can participate in as full a manner as is required of it from within and without. What does Kerr mean by "pure creative effort"?

Perhaps Kerr means the wasteful activity of artists and technoscientists who experiment without a prescribed goal in mind. Does Kerr endorse waste? This question will be ridiculed by those outside the academy (and perhaps by some within it) because of its promotion of waste under conditions of scarcity. Critics may even phrase their discontent in terms of the elitist attitude cherished by academics while still using the rhetoric of egalitarianism. Kerr is aware of the age-old predicament of the university, torn between an egalitarian commitment and the culture of merit and excellence to which only few are invited.

> The great university is of necessity elitist—the elite of merit—but it operates in an environment dedicated to an egalitarian philosophy. How may the contribution of the elite be made clear to the egalitarians, and how may an aristocracy of intellect justify itself to a democracy of all men? (Ibid., 121)

In my view, all students, however ill prepared, should be invited into the multiversity. They can find different tracks adjustable to their needs, interests, qualification, and aspirations. They can stay as long as they wish or leave at any time. They should be participants in determining the direction the academy takes, while acknowledging their dependence on their instructors, those whose prior knowledge and experience privileges them temporarily and for particular purposes under specific circumstances. Perhaps there is a way to break down the medieval hold on the university as it transforms itself into a multiversity; perhaps it can overthrow its indebtedness to the church and to its own hierarchical structure. Otherwise, its medieval garb will be changed only so slightly that the new garb will feel and appear no different, except of course for its pretense to be postmodern and not premodern. Will the faculty of the academy differ from their predecessors?

On the one hand, those involved in the academy are dupes of hegemonic culture—they do not even realize how they serve capitalist modes of production, distribution, and consumption. The subtleties of hegemonic socialization (toward efficiency claims, for example) may be perceived in the way in which the academy counts positions in departments and student hours per course, or how faculty performance is quantified, reported, and acted upon. On the other hand, those involved in the academy are facilitators of its reproduction, that is, active and willing participants. They champion the causes and goals of the academy, sometimes as a multiversity, with zeal and passion.

As far as Carl Boggs is concerned, what is at issue is the transformation of so-called liberal ideals of the academy that date back to the Enlightenment legacy concerning the autonomy of the university as a quasi-aristocratic site where even the church's influence is minimized to the ideals of postindustrial capitalism. Do academics in fact aspire to the ideals of virtue, justice, and reason, or are they merely the conveyers of the dominant culture's values of greed, self-interest, and competitiveness?

Are we to expect from academics any more than we expect from other cultural leaders who benefit directly from the fruits and spoils of capitalism? (Boggs 1993, 98ff) Boggs's critique of Kerr's notion of the multiversity is perhaps explicable in terms of the prescriptive nature of Kerr's characterization of the multiversity, and not in terms of its descriptive and explanatory power. For Kerr is correct in illustrating the extent to which the academy has transformed into a knowledge-production plant, where information flows are controlled and analyzed in accordance with the needs of the military-industrial complex or with the cultural concerns espoused by advertisers in the popular media. The academy is no longer an extension of the Catholic monastery turned into an aristocratic estate for the elite; nor is it the site for radical fermentation of alternative ideas and strategic alliances. Instead, says Boggs,

> Kerr's thesis embodied perhaps the clearest expression of the intellectual-cum-technocratic professional, and he articulated it from the vantage point of directing the largest and richest university complex in the world. (Ibid., 109)

Whether dupes or active perpetrators, whether performing the directives of corporate officers or those they deem their own (but only because they have internalized these directives, as Gideon Kunda (1992) argues), Boggs admonishes intellectuals in the academy for missing a critical dimension expected of them, a dimension one may call normative. It is in this respect, then, that I return to the question of intellectual responsibility.

If the multiversity is characterized by the knowledge industry, and if the knowledge industry *qua* industry is driven by so-called market forces, then is there room for pure creativity, for creativity for the sake of creativity? If art for art's sake and postmodern technoscience for postmodern technoscience's sake make sense, then creativity for creativity's sake can make cultural sense. In order for it to make sense, though, I would recommend that the academy must serve as a refuge for creative people, artists and humanists, technoscientists and intellectuals. As a refuge, this site can use its resources, its good will with government agencies and businesses, its prestige and credibility with the public at large, and its economies of scale for waste. Let me explain the last phrase.

In economic discourse, economies of scale is a situation wherein the sheer size of a plant producing a large number of products or services allows certain savings. For example, it is possible to save money on large orders and inventories, consolidate shipping, and reduce the size of middle management. Economies of scale is a term used to emphasize efficiency and productivity, cost-cutting and profitability. Assume that the concept makes sense and that its application is well known. Assume also that the academy operates not in order to maximize profits but rather to optimize its resources, both material and intellectual. Is it then not reasonable to argue that the academy is an optimal site for creative efforts the fruits of which remain obscure? Furthermore, is it not the perfect refuge from the harsh demands of short-term and short-sighted corporations? Finally, is the multiversity not the only site where waste can be tolerated for the sake of enhancing creativity?

I ask these questions rhetorically, since I believe that in the postmodern age of technoscience we need more, not less creativity. We need an active imagination with which to build a future better than our present and with which to avoid the horrors of the past; an imagination that would enable us to become more responsible for ourselves, our fellow humans, and the environment as a whole. Creativity, imagination, and art are all necessary means with which to confront our future. These activities may not pay off, nor will they necessarily attract a multitude of individuals. Instead, it may seem to administrators, legislators, and the public at large that there is a great deal of waste in the academy. So be it! Society as a whole cannot afford not to waste resources on its well-being and future, and thereby needs to tend to the academic soul. So, to promote waste in the academy is to promote the responsibility of academics and students for themselves and for the rest of the world.

Waste can take different forms, but when it is done deliberately for particular purposes, such as preserving the integrity of the academy or the future of society, it may be less obscure or counterintuitive. The concern with waste is a concern not only to defy contemporary cultural trends, but also to recall age-old concerns with how technoscientific innovations effect one's thinking and quality of life. For instance, Werner Heisenberg quotes the Chinese sage Chuang-Tzu:

> As Tzu-Gung was travelling through the regions north of the river Han, he saw an old man working in his vegetable garden. He had dug an irrigation ditch. The man would descend into a well, fetch up a vessel of water in his arms and pour it into the ditch. While his efforts were tremendous the results appeared to be very meager.
>
> Tzu-Gung said, "There is a way whereby you can irrigate a hundred ditches in one day, and whereby you can do much with little effort. Would you not like to hear of it?" Then the gardener stood up, looked at him and said, "And what would that be?"
>
> Tzu-Gung replied, "You take a wooden lever, weighted at the back and light in front. In this way you can bring up water so quickly that it just gushes out. This is called a draw-well."
>
> Then the anger rose up in the old man's face, and he said, "I have heard my teacher say that whoever uses machines does all his work like a machine. He who does his work like a machine grows a heart like a machine, and he who carries the heart of a machine in his breast loses his simplicity. He who has lost his simplicity becomes unsure in the strivings of his soul. Uncertainty in the strivings of the soul is something which does not agree with honest sense. It is not that I do not know of such things; I am ashamed to use them. (Heisenberg 1958, 20–21)

This short story is as much about the refusal to use technological innovations as about the "uncertainty in the strivings of the soul." In my view this is not only a story about the nostalgia for a world without machines and the spiritual value of manual labor, but a story that problematizes the desire for efficiency and the elimination of wasteful practices. The old man wants to spend his time irrigating his field in a manner that preserves his sense of dignity, and such an activity he does not find wasteful. The interests, pursuits, and practices of many intellectuals, artists, and academics may be deemed wasteful, but in defending this sort of waste I am pleading for human dignity and honor, however old-fashioned and privileged such a plea may sound in the age of postmodern technoscience.

It is also a plea to allow our imagination to catch up with the rapid changes of the time, as Anders argues. In Heisenberg's words: "In contradistinction to previous centuries this rapid change simply did not leave humanity time to get used to new conditions of life" (Ibid., 22). Perhaps the kind of waste I advocate here is a form of resistance against the ever-increasing encroachment of capitalist instrumental rationality.

As I bring this book to a close, I must remind myself that my journey is incomplete, that I will forever carry a promissory note concerning the question of personal and public responsibility. As a foreigner who transmits information from one language game to another, from one site to another, from one constituency to another, I have to remember the curse that shackles all foreigners: they remain foreigners even if invited and no matter how comfortable they may temporarily feel. This is also the postmodern condition that separates the temporal from the permanent, that allows one to claim privilege only within definite parameters or contexts and never for eternity or because of prior or everlasting authority. My personal account parallels what I see to be the postmodern technoscientific conditions of the transitionary moment between our century and the next, and perhaps one needs to personalize one's condition to be one with the world one inhabits. Perhaps that is what Spinoza meant when he translated the traditional concept of God to nature. By contrast, it may simply be a comfort zone one enjoys as the complexities of life keep on informing one's behavior and choices. Either way, there is no substitute for personal responsibility in the face of uncertainty and utter confusion. If anything, there is added responsibility to be honest about one's intentions, expectations, and plans.

Let me end with a quote from medieval Paris, where Philippe Auguste, the king of France, is supposed to have said the following of his own university faculty:

> They are hardier than knights. Knights, covered with their armour, hesitate to engage in battle. These clerics, who have neither hauberk nor helmet, but a tonsured head, playfully fall upon one another with daggers; most foolish behaviour, and very dangerous. (Schachner 1962, 341)

If we are bound to use daggers, we must exact care; if we are to use our tongues as daggers, we must exact special care; and if we are to use language responsibly, we had better be convinced that the battlefield is the last, not the first resort for resolving disagreement and displaying discontent.

References

Agassi, Joseph (1971), *Faraday as a Natural Philosopher*. Chicago: University of Chicago Press.

——— (1975), *Science in Flux*. Dordrecht and Boston: D. Reidel Publishing Co.

——— (1981), *Science and Society: Studies in the Sociology of Science*. Dordrecht: D. Reidel Publishing Co.

——— (1988), "The Future of Big Science," *Journal of Applied Philosophy* Vol. 5, 1:17–26.

——— (1991), "Pluralism and Science," *Methodology and Science* 24: 99–118.

——— and I.C. Jarvie, eds. (1987), *Rationality: The Critical View*. Dordrecht: Martinus Nijhoff.

Albritton, Robert (1993), "Marxian Political Economy for an Age of Postmodern Excess," *Rethinking Marxism* Vol. 6, 1:24–43.

Aronowitz, Stanley (1988), *Science as Power: Discourse and Ideology in Modern Science*. Minneapolis: University of Minnesota Press.

Ayer, A. J., ed. (1959), *Logical Positivism*. New York: The Free Press.

Bacon, Francis (1985), *The New Organon* [1620]. New York: Macmillan Publishing Co.

Ben-David, Joseph (1984), *The Scientist's Role in Society* [1971]. Chicago and London: University of Chicago Press.

Bergstrom, Lars (1993), "Quine, Underdetermination, and Skepticism," *The Journal of Philosophy* Vol. 90, 7: 331–358.

Best, Steven and Douglas Kellner (1991), *Postmodern Theory: Critical Interrogations*. New York: The Guilford Press.

Bloor, David (1991), *Knowledge and Social Imagery*. Chicago and London: University of Chicago Press.

Boggs, Carl (1993), *Intellectuals and the Crisis of Modernity*. Albany, NY: State University of New York Press.

Bohman, James (1991), *New Philosophy of Social Science: Problems of Indeterminacy*. Cambridge, MA: MIT Press.

Boland, Lawrence A. (1989), *The Methodology of Economic Model Building*. London: Routledge.

Bourdieu, Pierre (1988), *Homo Academicus* [1984], trans. P. Collier. Stanford: Stanford University Press.

——— (1990), *In Other Words: Essays Toward a Reflexive Sociology* [1987], trans. M. Adamson. Stanford: Stanford University Press.

——— and Loic J.D. Wacquant (1992), *An Invitation to Reflexive Sociology*. Chicago: University of Chicago Press.

Broad, William J. (1983), "Physicists Compete for the Biggest Project of All," *New York Times* September 20: C1, C8.

——— (1987), "Atom-Smashing Now and in the Future," *New York Times* February 3: C1, C4.

——— (1994), "Big Project Dead, Physicists Find Small Is Beautiful," *New York Times* April 19: B5, B9.

Browne, Malcolm W. (1990), "Supercollider's Rising Cost Provokes Opposition," *New York Times* May 29: C1, C6.

——— (1990), "Soviets to Contribute $200 Million to Plan on U.S. Accelerator," *New York Times* October 26: A16.

——— (1992), "Physicists Struggle to Save Supercollider from Budgetary Ax," *New York Times* July 7: B7, B9.

Bunge, Mario (1991), "Five Bridges Between Scientific Disciplines," in *The Cybernetics of Complex Systems: Self-Organization, Evolution, and Social Change*, ed. Felix Geyer. Salinas, CA: Intersystems Publications, pp. 1–10.

Burtt, E. A. (1982), *The Metaphysical Foundations of Modern Science* [1924]. Atlantic Highlands, NJ: Humanities Press.

Buskirk, Martha (1992), "Commodification as Censor: Copyrights and Fair Use," *October* 60: 83–109.

Byrne, John A. (1993), "The Virtual Corporation," *Business Week* February 8: 98–103.

Callinicos, Alex (1990), *Against Postmodernism: A Marxist Critique.* New York: St. Martin's Press.

Campbell, Donald T. (1993), "Plausible Coselection of Belief by Referent: All the 'Objectivity' that is Possible," *Perspectives on Science*, Vol. 1, 1: 88–108.

Cohen, I. Bernard (1985), *Revolution in Science.* Cambridge, MA: Harvard University Press.

Collingwood, R. G. (1958), *The Principles of Art* [1938]. London and Oxford: Oxford University Press.

Collins, Randall (1993), "Ethical Controversies of Science and Society: A Relation Between Two Spheres of Social Conflict,"*Controversial Science: From Content to Contention*, ed. Thomas Brante, Steve Fuller, and William Lynch. Albany, NY: State University of New York Press, pp. 301–317.

Congress (1985a), "International Cooperation in *Big Science*: High Energy Physics." Committee on Science and Technology, House of Representatives; Science Policy Study, Hearing Volume 4; Ninety-Ninth Congress, First Session (April 25).

Congress (1985b), "International Cooperation in Science." Committee on Science and Technology, House of Representatives; Science Policy Study, Hearing Volume 7; Ninety-Ninth Congress, First Session (June 18–20, 27).

Connor, Steven (1989), *Postmodernist Culture: An Introduction to Theories of the Contemporary.* Oxford: Basil Blackwell.

Cordes, Colleen (1993), "Universities Angered by U.S. Proposal to Deny Them Patents on Some Research," *The Chronicle of Higher Education* September 22: A26, A28.

Davidson, Donald (1984), *Inquiries into Truth and Interpretation.* Oxford: Clarendon Press.

de Beauvoir, Simone (1991), *The Ethics of Ambiguity* [1948], trans. B. Frechtman. New York: Citadel Press.

Derrida, Jacques (1988), *Limited Inc* [1977], trans. S. Weber and J. Mehlman. Evanston, IL: Northwestern University Press.

Duhem, Pierre (1969), *To Save the Phenomena: An Essay on the Idea of Physical Theory from Plato to Galileo* [1908], trans. E. Dolan and C. Maschler. Chicago and London: University of Chicago Press.

——— (1991), *German Science* [1915], trans. J. Lyon. La Salle, IL: Open Court.

Eatherly, Claude and Gunther Anders (1989), *Burning Conscience: The Guilt of Hiroshima* [1961]. New York: Paragon House.

Eco, Umberto (1994), *Apocalypse Postponed*. Bloomington and Indianapolis: Indiana University Press.

Einstein, Albert (1954), *Ideas and Opinions*, trans. S. Bargmann. New York: Bonanza Books.

Engels, Frederick (1894), *Herr Eugen Dühring's Revolution in Science* [1878]. New York: International Publishers.

Faust, David (1984), *The Limits of Scientific Reasoning*. Minneapolis: University of Minnesota Press.

Feyerabend, Paul (1975), *Against Method: Outline of an Anarchistic Theory of Knowledge*. London: Verso.

Fleck, Ludwik (1979), *The Genesis and Development of a Scientific Fact* [1935], trans. F. Bradley and T.J. Trenn. Chicago and London: University of Chicago Press.

Forman, Paul (1971), "Weimar Culture, Causality, and Quantum Theory, 1918–1927: Adaptation by German Physicists and Mathematicians to a Hostile Intellectual Environment," *Historical Studies in the Physical Sciences*, ed. Russell McCormmach. Philadelphia: University of Pennsylvania Press.

Foucault, Michel (1970), *The Order of Things: An Archeology of the Human Sciences* [1966]. New York: Vintage Books.

———— (1979), *Discipline & Punish: The Birth of the Prison* [1975], trans. A. Sheridan. New York: Vintage Books.

Franklin, Ben A. (1987), "Reagan to Press for $6 Billion Atom Smasher," *New York Times* January 31: A1, A10.

——— (1988), "Texas is Awarded Giant U.S. Project on Smashing Atom," *New York Times* November 11: A1, A25.

Friedman, Milton (1982), *Capitalism and Freedom* [1962]. Chicago and London: University of Chicago Press.

Fuller, Steve (1988), *Social Epistemology*. Bloomington and Indianapolis: Indiana University Press.

Gatens-Robinson, Eugenie (1993), "Why Falsification is the Wrong Paradigm for Evolutionary Epistemology: An Analysis of Hull's Selection Theory," *Philosophy of Science* 60:535–557.

Gellner, Ernest (1992), *Postmodernism, Reason and Religion*. London and New York: Routledge.

Gleick, James (1987), "Advances Pose Obstacle to Atom Smasher Plan," *New York Times* April 14: C1, C6.

Goonatilake, Susantha (1993), "Modern Science and the Periphery: The Characteristics of Dependent Knowledge," in *The Racial Economy of Science*, ed. Sandra Harding. Bloomington and Indianapolis: Indiana University Press, pp. 259–267.

Greenberg, Valerie D. (1990), *Transgressive Readings: The Texts of Franz Kafka and Max Plank*. Ann Arbor, MI: University of Michigan Press.

Guattari, Felix (1986), "The Postmodern Dead End," *Flash Art* 128:40–41.

Habermas, Jürgen (1979), *Communication and the Evolution of Society* [1976], trans. T. McCarthy. Boston: Beacon Press.

Hanson, Norwood Russell (1958), *Patterns of Discovery: An Inquiry into the Conceptual Foundations of Science*. Cambridge: Cambridge University Press.

Haraway, Donna J. (1991), *Simians, Cyborgs, and Women: The Reinvention of Nature*. New York: Routledge; especially " 'Gender' for a Marxist Dictionary: The Sexual Politics of a Word" originally published 1987; "A Cyborg Manifesto: Science, Technology, and Socialist-Feminism in the Late Twentieth Century" originally published 1985; and "Situated Knowledges: The Science Question in Feminism and the Privilege of Partial Perspective" originally published 1988.

——— (1992), "Otherworldly Conversations; Terran Topics; Local Terms," *Science as Culture* Vol. 3, 14: 64–98.

Harding, Sandra (1986), *The Science Question in Feminism*. Ithaca and London: Cornell University Press.

——— (1991), *Whose Science? Whose Knowledge?* Ithaca: Cornell University Press.

Harvey, David (1989), *The Condition of Postmodernity: An Enquiry into the Origins of Cultural Change*. London: Basil Blackwell.

Hayles, N. Katherine, ed. (1991), *Chaos and Order: Complex Dynamics in Literature and Science*. Chicago and London: University of Chicago Press.

Heisenberg, Werner (1958), *The Physicist's Conception of Nature*, trans. A.J. Pomerans. Westport, CT: Greenwood Press.

Hodgson, Geoffrey M. (1992), "The Reconstruction of Economics: Is There Still a Place for Neoclassical Theory?," *Journal of Economic Issues* Vol. 26, No. 3, 749–767.

Hollis, Martin and Edward Nell (1975), *Rational Economic Man: A Philosophical Critique of Neo-Classical Economics*. London and New York: Cambridge University Press.

Hoesterey, Ingeborg, ed. (1991), *Zeitgeist in Babel: The Postmodernist Controversy*. Bloomington: Indiana University Press.

Hutcheon, Linda (1989), *The Politics of Postmodernism*. London and New York: Routledge.

Jameson, Fredric (1991), *Postmodernism, or, The Cultural Logic of Late Capitalism*. Durham, NC: Duke University Press.

Jarvie, I. C. (1972), *Concepts and Society*. London and Boston: Routledge & Kegan Paul.

Kantorovich, Aharon (1993), *Scientific Discovery: Logic and Tinkering*. Albany, NY: State University of New York Press.

Keller, Evelyn Fox (1985), *Reflections on Gender and Science*. New Haven and London: Yale University Press.

Kerr, Clark (1963), *The Uses of the University*. Cambridge, MA: Harvard University Press.

Kirk, G. S., J. E. Raven, and M. Schofield, eds. (1983), *The Presocratic Philosophers* [1957]. Cambridge: Cambridge University Press.

Krauss, Clifford (1992), "Supercollider Regaining Its Support in Congress," *New York Times* July 25: A10.

——— (1993), "Deficit Taking Toll on Lawmakers' Dreams on Big-Science Projects," *New York Times* June 28: A10.

Kuhn, Thomas S. (1970), *The Structure of Scientific Revolutions* [1962]. Chicago: University of Chicago Press.

Kunda, Gideon (1992), *Engineering Culture: Control and Commitment in a High-Tech Corporation*. Philadelphia: Temple University Press.

Lakatos, Imre (1970), "Falsification and the Methodology of Scientific Research Programmes," in *Criticism and the Growth of Knowledge*, ed. Imre Lakatos and Alan Musgrave. Cambridge: Cambridge University Press, pp. 91–196.

——— (1976), *Proofs and Refutations: The Logic of Mathematical Discovery*. London and New York: Cambridge University Press.

Laor, Nathaniel (1990), "Seduction in Tongues: Reconstructing the Field of Metaphor in the Treatment of Schizophrenia," *Prescriptions: The Dissemination of Medical Authority*, ed. Gayle L. Ormiston and Raphael Sassower. Westport, CT: Greenwood Press, pp. 141–175.

Latour, Bruno (1987), *Science in Action: How to Follow Scientists and Engineers through Society*. Cambridge: Harvard University Press.

——— (1990), "Postmodern? No, Simply Amodern! Steps Towards an Anthropology of Science," *Studies in the History and Philosophy of Science* 21: 145–171.

——— (1993), *We Have Never Been Modern* [1991], trans. C. Porter. Cambridge: Harvard University Press.

——— and Steve Woolgar (1986), *Laboratory Life: The Construction of Scientific Facts* [1979]. Princeton, NJ: Princeton University Press.

Laudan, Larry (1990), *Science and Relativism*. Chicago and London: University of Chicago Press.

Locke, John (1964), *An Essay Concerning Human Understanding* [1690]. New York: New American Library.

Longino, Helen A. (1990), "Feminism and Philosophy of Science," *Journal of Social Philosophy*, Vol 21, Nos 2–3:150–159.

Lyotard, Jean-François (1984), *The Postmodern Condition: A Report on Knowledge* [1979], trans. G. Bennington and B. Massumi. Minneapolis: University of Minnesota Press.

——— (1988a), *The Differend: Phrases in Dispute* [1983], trans. G. Van Den Abbeele. Minneapolis: University of Minnesota Press.

——— (1988b), *Peregrinations: Law, Form, Event*. New York: Columbia University Press.

——— (1991), *The Inhuman: Reflections on Time* [1988], trans. G. Bennington and R. Bowlby. Stanford: Stanford University Press.

——— (1993), *Political Writings*, trans. B. Readings and K. Paul. Minneapolis: University of Minnesota Press.

——— and Jean-Loup Thèbaud (1985), *Just Gaming* [1979], trans. W. Godzich. Minneapolis: University of Minnesota Press.

Machlup, Fritz (1962), *The Production and Distribution of Knowledge in the United States*. Princeton, NJ: Princeton University Press.

Marcuse, Herbert (1964), *One Dimensional Man*. Boston: Beacon Press.

Marx, Karl & Frederick Engels (1988), *The German Ideology* [1846] (Selections). New York: International Publishers.

Merton, Robert K. (1968), *Social Theory and Social Structure* [1949]. New York: Free Press.

——— (1985), *On the Shoulders of Giants* [1965]. Chicago and London: University of Chicago Press.

Morrell, Jack and Arnold Thackray (1981), *Gentlemen of Science: Early Years of the British Association for the Advancement of Science*. Oxford: Clarendon Press.

Murphy, Nancey (1990), "Scientific Realism and Postmodern Philosophy," *British Journal of Philosophy of Science* 41: 291–303.

Myrdal, Gunnar (1969), *Objectivity in Social Research*. New York: Pantheon Books.

Newman, Charles (1985), *The Post-Modern Aura: The Act of Fiction in an Age of Inflation*. Evanston, IL: Northwestern University Press.

Nicolson, Linda J., ed. (1990), *Feminism/Postmodernism*. New York and London: Routledge.

Nietzsche, Friedrich (1974), *The Gay Science* [1882/7], trans. W. Kaufmann. New York: Vintage.

Norris, Christopher (1990), *What's Wrong with Postmodernism: Critical Theory and the Ends of Philosophy*. Baltimore: Johns Hopkins University Press.

Nussbaum, Martha (1994), "Feminism and Philosophy," *New York Review of Books* October 20: 59–63.

Ormiston, Gayle and Raphael Sassower (1989), *Narrative Experiments: The Discursive Authority of Science and Technology*. Minneapolis: University of Minnesota Press.

——— (1991), "Interpretive Displacements and Seductions of Pluralism," *Social Epistemology* 5: 311–316.

——— (1993), "From Marx's Politics to Rorty's Poetics: Shifts in the Critique of Metaphysics," *Man and World* 26: 63–82.

Parusnikova, Zuzana (1992), "Is a Postmodern Philosophy of Science Possible?," *Studies in the History and Philosophy of Science* 23: 21–37.

Pickering, Andy (1990), "Openness and Closure: On the Goals of Scientific Practice," in *Experimental Inquiries*, ed. H.E. Le Grand. Dordrecht: Kluwer, pp. 215–239.

Polanyi, Karl (1957), *The Great Transformation: The Political and Economic Origins of Our Time* [1944]. Boston: Beacon Press.

Polanyi, Michael (1958), *Personal Knowledge: Towards a Post-Critical Philosophy*. New York: Harper & Row.

Popper, Karl R. (1957), *The Poverty of Historicism*. New York: Harper & Row.

——— (1959), *The Logic of Scientific Discovery* [1935]. New York: Harper & Row.

——— (1963), *Conjectures and Refutations: The Growth of Scientific Knowledge*. New York: Harper & Row.

——— (1966), *The Open Society and Its Enemies*, Vol. 2: "Hegel and Marx," revised ed. Princeton, NJ: Princeton University Press.

——— (1979), *Objective Knowledge: An Evolutionary Approach*, revised ed. Oxford: Clarendon Press.

Presley, C. F. (1967), "Quine," *The Encyclopedia of Philosophy* Vol. 7. New York: Macmillan Publishing, pp. 53–55.

Quine, Willard Van Orman (1963), *From a Logical Point of View* [1953]. New York: Harper & Row.

——— (1960), *Word & Object*. Cambridge, MA: MIT Press.

——— (1969), *Ontological Relativity and Other Essays*. New York and London: Columbia University Press.

Radnitzky, Gerard (1991), "Refined Falsificationism Meets the Challenge from the Relativist Philosophy of Science," *British Journal of Philosophy of Science* 42: 273–284.

Remington, John A. (1988), "Beyond Big Science in America: The Binding of Inquiry," *Social Studies of Science* 18: 45–72.

Rorty, Richard (1989), *Contingency, Irony, and Solidarity*. Cambridge: Cambridge University Press.

——— (1991), *Objectivity, Relativism, and Truth: Philosophical Papers*, Vol. I. Cambridge: Cambridge University Press.

Rosenau, Pauline Marie (1992), *Post-Modernism and the Social Sciences*. Princeton: Princeton University Press.

Rosenstein-Rodan, Paul N. (1958), "Problems of Industrialization of Eastern and South-Eastern Europe" [1943], *The Economics of Underdevelopment*, ed. A. N. Agarwala and S. P. Singh. London and Oxford, Oxford University Press, pp. 245–255.

Rosenthal, Andrew (1992), "White House Fight on Collider Deal," *New York Times* September 26: A1, A9.

Rouse, Joseph (1991), "The Politics of Postmodern Philosophy of Science," *Philosophy of Science* 58: 607–627.

Sanger, David E. (1990), "U.S. Asks Japanese to Join in Project for Supercollider," *New York Times* June 1: A1, D16.

——— (1991), "U.S. Offers Japan Major Role in Bid to Save Supercollider Project," *New York Times* October 16: A14.

Sassower, Raphael (1988), "Ideology Masked as Science: Shielding Economics from Criticism," *Journal of Economic Issues* Vol. 22, 1:167–179.

——— (1990a), "Scarcity and Setting the Boundaries of Political Economy," *Social Epistemology* Vol. 4, 1: 75–91.

——— (1990b), "Medical Education: The Training of Ethical Physicians," *Studies in Philosophy and Education* Vol. 10, 1: 251–261.

——— (1993a), *Knowledge Without Expertise: On the Status of Scientists*. Albany, NY: State University of New York Press.

——— (1993b), "Postmodernism and Philosophy of Science," *Philosophy of the Social Sciences* Vol. 23, 4: 426–445.

——— (1994a), "The Politics of Situating Knowledge: An Exercise in Social Epistemology," *Argumentation* 8: 185–198.

——— (1994b), "Verlighetens ofrankomlighet" (in Swedish), ["The Inevitability of Reality"], *Vest*, Vol. 7, 3: 21–36.

——— and Joseph Agassi (1994), "Avoiding the Posts," *Critical Review* Vol. 8, 1: 95–111.

——— and Charla P. Ogaz (1991), "Philosophical Hierarchies and Lyotard's Dichotomies," *Philosophy Today* 35: 153–160.

Schachner, Nathan (1962), *The Mediaeval Universities* [1938]. New York: A.S. Barnes & Co.

Schumacher, E. F. (1973), *Small is Beautiful: Economics as if People Mattered*. New York: Harper & Row.

Serres, Michel (1989), *Detachment* [1986], trans. G. James and R. Federman. Athens: Ohio University Press.

Sherman, Howard J. (1993), "The Relational Approach to Political Economy," *Rethinking Marxism* Vol. 6, 4: 104–116.

Snow, C. P. (1964), *The Two Cultures and A Second Look* [1959]. Cambridge and New York: Cambridge University Press.

Stanley, Manfred (1978), *The Technological Conscience: Survival and Dignity in an Age of Expertise*. Chicago and London: University of Chicago Press.

Toulmin, Stephen (1981), "The Emergence of Post-Modern Science," *The Great Ideas Today*. Chicago: Encyclopedia Britannica, pp. 68–114.

——— (1985), "Pluralism and Responsibility in Post-Modern Science," *Science, Technology & Human Values*, 10: 28–37.

Winch, Peter (1964), "Understanding a Primitive Society," *American Philosophical Quarterly* Vol. 1, 4: 307–324.

Wines, Michael (1993), "House Kills the Supercollider, And Now It Might Stay Dead," *New York Times* October 20: A1, A20.

Wittgenstein, Ludwig (1968), *Philosophical Investigations* [1958]. New York: Macmillan.

Index